Chemistry Data Book
Second Edition in SI

Compiled by

J. G. Stark M.A. (Cantab.)
Glasgow Academy

H. G. Wallace M.A. (Cantab.), Dip.Ed.
Carmel School, Darlington

Foreword by

M. L. McGlashan Ph.D., D.Sc., C.Chem., F.R.S.C.
Professor of Chemistry at University College, London
Former Chairman of the Commission on Symbols, Terminology and
Units of the Division of Physical Chemistry, I.U.P.A.C.

John Murray 50 Albemarle Street London

© J. G. Stark, H. G. Wallace, 1969, 1970, 1982
First edition 1969
SI edition 1970
Second edition in SI 1982
Reprinted 1984, 1986, 1988, 1990, 1991, 1992, 1994, 1995,
1996, 1997, 1999, 2000, 2001, 2003

Published by John Murray (Publishers) Ltd, a member of the Hodder Headline Group
338 Euston Road, London NW1 3BH

Printed and bound in Great Britain by
Athenæum Press Ltd, Gateshead, Tyne & Wear

British Library Cataloguing in Publication Data

Stark, J. G.
   Chemistry data book.–2nd ed.
   1. Chemical elements–Tables
   2. Chemistry–Tables
   I. Title II. Wallace, H. G.
   540′212   QD31.2
   ISBN 0–7195–3951–X

Preface

Since the SI edition of this book was published in 1970, the chemical nomenclature (particularly for organic compounds) has been extensively revised, minor changes have been made to the data in various tables and an index has been added. In this new edition the opportunity has been taken to correct some remaining errors, update the data and delete a number of tables which are no longer relevant to the majority of A-level syllabuses.

The following tables have been added for this edition:

Equilibrium constants for gaseous reactions at various temperatures

Electrolytic conductivity of water at various temperatures

Infrared data.

In addition Gibbs free energies of formation of aqueous ions have been included in Table 32 and the table of lattice enthalpies has been extended to include theoretical values.

The use of data in the teaching of chemistry has become increasingly important in recent years and the authors' intention remains to provide a book which will be used as a comprehensive source of the chemical data which are likely to be needed by sixth-form and technical college students and first-year university students. It is hoped that the book will be used not only for reference, but also in the discussion of problems and ideas arising from practical work. The general layout of the tables is as follows:

physical constants and fundamental particles
the elements
basic atomic and molecular properties
thermodynamic properties
kinetic data
organic data
analytical data
miscellaneous data.

The detailed list of contents should enable the reader to find the data he or she needs quite easily. There is also an index at the end of the book. Many of the tables are preceded by a brief explanation of the data contained in them.

The compilers would welcome any constructive criticisms and suggestions from readers which will increase the usefulness of the book.

August 1982   J. G. Stark
                                                                H. G. Wallace

Acknowledgements

The authors wish to acknowledge the many sources used in the compilation of this data book and particularly the following:

F. A. Cotton and G. Wilkinson, *Advanced Inorganic Chemistry* (Interscience, New York, 2nd edn., 1966).

Chemical Bond Approach Committee, *Chemical Systems* (McGraw-Hill, New York, 1964).

K. B. Harvey and G. B. Porter, *Physical Inorganic Chemistry* (Addison-Wesley, Reading, Mass., 1963).

W. M. Latimer, *Oxidation Potentials* (Prentice-Hall, Englewood Cliffs, New Jersey, 2nd edn., 1952).

T. Moeller, *Inorganic Chemistry* (Wiley, New York, 1952).

Periodic Table of the Elements and *Table of Periodic Properties of the Elements* (E. H. Sargent and Co.).

R. C. Weast (ed.), *Handbook of Chemistry and Physics* (Chemical Rubber Co., 47th edn., 1966).

They are also grateful to Professor M. L. McGlashan for his advice and criticism and for writing a foreword to the book and to Dr A. D. Pethybridge, of the University of Reading, for his helpful suggestions in relation to Table 1.

Foreword

The metric system of units had 'just grown', since the Metre Convention of 1875, until it contained many redundant units. For example the joule, the erg, the kilogram-force metre, the kilowatt hour, the electronvolt, at least three calories differing from one another by as much as one part in a thousand, the thermie, and the litre atmosphere, were all used as units of energy in 'metric' countries and by scientists everywhere. To these, Anglo-Saxon technologists added the poundal, the foot pound-force, the British thermal unit, and the horsepower hour.

In 1960 a small but sufficient set of metric units was carefully chosen by the General Conference on Weights and Measures and given the name *SI Units*, and these together with their decimal multiples and sub-multiples formed by use of the *SI Prefixes*, became the *International System of Units (SI)*. The adoption of the SI has been strongly urged on the scientific and technological world by the International Organization for Standardization and by the International Scientific Unions including that of Pure and Applied Chemistry, and it is in fact rapidly being adopted throughout the scientific literature. It has already been adopted in several major countries as the only legal system of weights and measures for purposes of trade, and is about to be adopted in many more.

If the United Kingdom were already 'metric' (as we so nearly were in 1871 when a Bill for the compulsory adoption of the metric system was defeated by only five votes on its Second Reading in the House of Commons) then we should no doubt now be in the process of 'going SI'. By luck the SI was ready for us when in 1965 we were at last ready to take the decision to 'go metric'. According to the Government's Metrication programme 1975 was the deadline by which the SI should have been adopted throughout the Nation's industry, commerce, and daily life. That target was not reached but nevertheless steady progress is being made.

Students of chemistry must learn to use the SI and, more importantly, to 'think SI', whatever their future careers may be.

Mr Stark and Mr Wallace have served chemists and chemistry well by producing their book, which contains a most useful collection of the results of physicochemically important measurements expressed as the numbers which are equal to the ratios of a physical quantity and an appropriate SI unit.

M. L. McGlashan

Contents

Preface		v
Acknowledgements		vi
Foreword by Professor M. L. McGlashan		vii
Thermodynamic sign convention		xii
SI units		xiii
1	Physical constants and conversion factors	1
2	Fundamental particles	1
3	The electromagnetic spectrum; the visible spectrum	2

The elements

4	The periodic classification (with atomic numbers)	3
5	Electronic configurations of the elements (ground states)	4
6	Energy levels for a many-electron atom	7
7	Properties of the elements: symbol, atomic number, relative atomic mass, density, melting temperature, boiling temperature, specific heat capacity, oxidation states	8
8	Crystal structures of the solid elements	12
9	Stable isotopes of the elements	13
10	Half-lives of radioactive elements	16
11	Radioactive series	17
12	Electric conductivities of the elements at 298 K	18

Basic atomic and molecular properties

13	Ionization energies of the elements: first ionization energies; successive ionization energies	19
14	Electron affinities of the elements	22
15	Atomic spectrum of hydrogen	23
16	Electronegativities of the elements (Pauling)	24
17	Percentage ionic character of a single bond	25
18	Dipole moments of inorganic compounds in the vapour phase	26
19	Metallic, covalent and van der Waals radii	27
20	Ionic radii	28
21	The shapes of molecules and ions	30

22 Spatial distribution of hybrid orbitals — 31
23 Covalent bond lengths — 31
24 Average bond enthalpies at 298 K — 32

Thermodynamic properties

25 Thermodynamic data for inorganic compounds, organic compounds: enthalpy of formation, Gibbs free energy of formation, entropy, heat capacity — 33
26 Thermodynamic data relating to change of state for elements, compounds: enthalpy of fusion, enthalpy of vaporization, enthalpy of atomization — 50
27 Thermodynamics of metal extraction — 53
28 Enthalpies of combustion — 54
29 Enthalpies of hydrogenation of gases — 55
30 Enthalpies of solution at 298 K — 56
31 Enthalpies of neutralization at 298 K — 56
32 Enthalpies and Gibbs free energies of formation of aqueous ions — 57
33 Hydration enthalpies of ions — 58
34 Lattice enthalpies at 298 K (experimental and theoretical values): alkali metal halides; other substances — 59
35 Solubility of gases in water — 60
36 Solubility of inorganic compounds — 62
37 Azeotropes: binary mixtures at atmospheric pressure — 64
38 Vapour pressure of water at various temperatures — 65
39 Eutectic mixtures — 66
40 Transition temperatures — 66
41 Typical equilibrium data — 67
42 Equilibrium constants for gaseous reactions at various temperatures — 68
43 Standard electrode potentials — 70
44 Standard reference electrodes — 73
45 Dissociation constants of inorganic acids — 74
46 Ionic product of water at various temperatures — 75
47 Solubility products — 76
48 Stability (or formation) constants of complex ions at 298 K — 77
49 Electrolytic conductivity of potassium chloride solution — 79

50 Molar conductivity of aqueous solutions at 298 K 80

51 Molar conductivity of ions at infinite dilution at 298 K 80

52 Electrolytic conductivity of water at various temperatures 81

Kinetic data

53 Kinetic data 82

54 Experimental rate laws 85

55 Activation energies 86

Organic data

56 Physical properties of organic compounds: melting temperature, boiling temperature, density, refractive index, solubility 87

57 Melting temperatures of organic derivatives 94

58 Strengths of organic acids 96

59 Dipole moments of organic compounds 100

Analytical data

60 Standard solutions for titrimetric analysis 102

61 Acid-base indicators 103

62 Standard buffer solutions 103

63 Laboratory reagents 104

64 Infrared data 104

Miscellaneous data

65 Composition of the atmosphere 105

66 Composition of the earth's crust 105

Logarithms and Antilogarithms 106

Index 111

Thermodynamic Sign Convention

All thermodynamic data in this book refer to the standard temperature of 298 K. The sign convention used is that ΔH denotes the difference between the final and the initial enthalpy, i.e. $\Delta H = H_f - H_i$.

SI Units

There are seven *basic SI units* as shown in Table I.

Table I Basic SI Units

Physical quantity	Name of unit	Symbol
Length	metre	m
Mass	kilogram	kg
Time	second	s
Electric current	ampere	A
Thermodynamic temperature	kelvin (*not* degree Kelvin)	K (*not* °K)
Luminous intensity	candela	cd
Amount of substance	mole	mol

In addition, there are *derived units* which are defined in terms of the basic units or other derived units. Some of those most likely to be met by the user of this book are shown in Table II.

Table II Derived Units

Physical quantity	Name of unit	Symbol and definition
Force	newton	$N\,(kg\,m\,s^{-2} = J\,m^{-1})$
Energy, heat	joule	$J\,(kg\,m^2\,s^{-2})$
Power	watt	$W\,(kg\,m^2\,s^{-3} = J\,s^{-1})$
Electric charge	coulomb	$C\,(A\,s)$
Electric potential difference	volt	$V\,(kg\,m^2\,s^{-3}\,A^{-1} = J\,A^{-1}\,s^{-1})$
Electric resistance	ohm	$\Omega\,(kg\,m^2\,s^{-3}\,A^{-2} = V\,A^{-1})$
Electric capacitance	farad	$F\,(A^2\,s^4\,kg^{-1}\,m^{-2} = A\,s\,V^{-1})$
Magnetic flux	weber	$Wb\,(kg\,m^2\,s^{-2}\,A^{-1} = V\,s)$
Magnetic flux density	tesla	$T\,(kg\,s^{-2}\,A^{-1} = V\,s\,m^{-2})$
Inductance	henry	$H\,(kg\,m^2\,s^{-2}\,A^{-2} = V\,s\,A^{-1})$
Frequency	hertz	$Hz\,(s^{-1})$

There are also derived units, such as those of volume, density, pressure and specific heat capacity, which are based on SI units. The names and symbols of these units are derived from those in Tables I and II as shown in Table III.

Table III   **Other Derived Units Based on SI**

Physical quantity	Name of unit	Symbol
Area	square metre	m^2
Volume	cubic metre	m^3
Velocity	metre per second	$m\ s^{-1}$
Acceleration	metre per second squared	$m\ s^{-2}$
Density	kilogram per cubic metre	$kg\ m^{-3}$
Pressure	newton per square metre (pascal)	$N\ m^{-2}$ (Pa) $= kg\ m^{-1}\ s^{-2}$
Surface tension	newton per metre	$N\ m^{-1} = kg\ s^{-2}$
Electric field strength	volt per metre	$V\ m^{-1} = kg\ m\ s^{-3}\ A^{-1}$
Magnetic field strength	ampere per metre	$A\ m^{-1}$
Dipole moment	coulomb metre	$C\ m = A\ s\ m$
Magnetic moment	ampere square metre	$A\ m^2$
Heat capacity ⎱ Entropy ⎰	joule per kelvin	$J\ K^{-1} = kg\ m^2\ s^{-2}\ K^{-1}$
Specific heat capacity	joule per kilogram kelvin	$J\ kg^{-1}\ K^{-1} = m^2\ s^{-2}\ K^{-1}$
Concentration	mole per cubic metre	$mol\ m^{-3}$

Multiples of units in powers of ten are indicated by means of agreed prefixes and symbols as shown in Table IV.

Table IV   **Multiples of Units**

Multiple	Prefix	Symbol	Multiple	Prefix	Symbol
10^{12}	tera	T	10^{-1}	deci	d
10^{9}	giga	G	10^{-2}	centi	c
10^{6}	mega	M	10^{-3}	milli	m
10^{3}	kilo	k	10^{-6}	micro	μ
10^{2}	hecto	h	10^{-9}	nano	n
10	deca	da	10^{-12}	pico	p

Details of SI units can be found in the following publications:

The International System (SI) Units (B.S. 3763), British Standards Institution, 1970.

M. L. McGlashan, *Physico-Chemical Quantities and Units*, 2nd edn, Royal Institute of Chemistry, 1971.

Quantities, Units, and Symbols, The Royal Society, 1971.

1 Physical Constants and Conversion Factors

Speed of light in a vacuum $(c) = 2.997\,925 \times 10^8$ m s^{-1}
Boltzmann constant $(k) = 1.380\,662 \times 10^{-23}$ J K^{-1}
Planck constant $(h) = 6.626\,176 \times 10^{-34}$ J s
Rydberg constant $(R_H) = 1.097\,373 \times 10^7$ m^{-1}
Avogadro constant $(L) = 6.022\,045 \times 10^{23}$ mol^{-1}
Gas constant $(R) = 8.314\,41$ J K^{-1} mol^{-1}
$\qquad\qquad\qquad = 1.987\,1$ cal K^{-1} mol^{-1}
$\qquad\qquad\qquad = 82.053$ cm^3 atm K^{-1} mol^{-1}
'Ice-point' temperature $(T_{ice}) = 273.150\,0$ K
Faraday constant $(F) = 9.648\,456 \times 10^4$ C mol^{-1}

Molar volume of an ideal gas at s.t.p.[a] $(V_m^{\ominus}) = 2.241\,383 \times 10^{-2}$ m^3 mol^{-1}
Triple point temperature of water $= 273.16$ K
Potential of a standard Weston cell (saturated) at θ_C
$\qquad\qquad\qquad\qquad = [1.018\,3 - 4 \times 10^{-5}(\theta_C/^{\circ}C - 20)]$V
Nernst factor $\left(\dfrac{RT\ln 10}{F}\right)$ at 25 $^{\circ}$C $= 0.059\,159$ V

1 calorie (cal) $= 4.184$ J
1 electrostatic unit of charge $\hat{=} 3.335\,640 \times 10^{-10}$ C
L eV[b] $\hat{=} 96.484$ kJ mol^{-1}
1 atmosphere (atm) $= 760$ torr (≈ 760 mmHg) $= 101\,325$ N m^{-2} (Pa)
1 ångström (Å) $= 10^{-10}$ m $- 10^{-8}$ cm $= 10^{-4}$ μm
$\qquad\qquad\qquad\qquad\qquad = 10^{-1}$ nm $= 10^2$ pm
1 litre (l) $= 10^{-3}$ m^3 $= 1$ dm^3 $= 10^3$ cm^3
1 curie (Ci) $= 3.7 \times 10^{10}$ s^{-1}
Conversion of Celsius temperature to thermodynamic temperature:
$\quad \theta_C/^{\circ}C = T/K - 273.150$
$\ln x = 2.303 \log_{10} x$

2 Fundamental Particles

	Proton	*Neutron*	*Electron*
Symbol	p	n	e
Mass	$1.672\,648 \times 10^{-27}$ kg	$1.674\,954 \times 10^{-27}$ kg	$9.109\,534 \times 10^{-31}$ kg
Charge	$1.602\,189 \times 10^{-19}$ C	0	$1.602\,189 \times 10^{-19}$ C
Mass relative to electron	1836	1839	1
Charge relative to proton	$+1$	0	-1

[a] That is 273.15 K and 101 325 N m^{-2} pressure.
[b] The quantity L eV is commonly, but incorrectly, known as an 'electronvolt'.

3 The Electromagnetic Spectrum

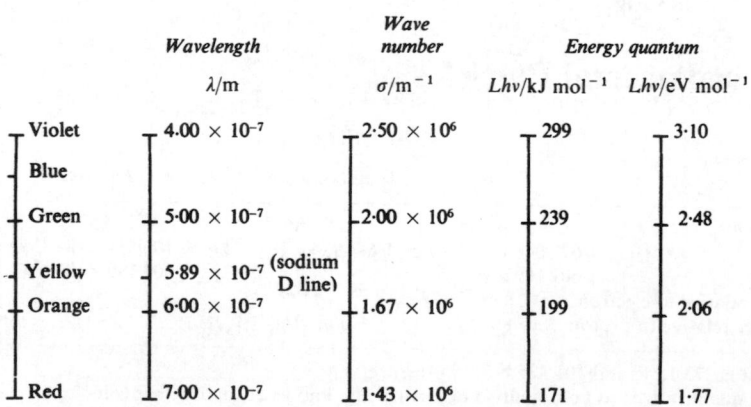

	Wavelength λ/m	Wavelength λ	Wave number σ/m^{-1}	Frequency ν/MHz	Energy quantum $Lh\nu/\mathrm{kJ\ mol^{-1}}$	$Lh\nu/\mathrm{eV\ mol^{-1}}$	Molecular phenomena
Gamma rays	10^{-12}	10^{-2} Å	10^{12}	$3{\cdot}00 \times 10^{14}$	$1{\cdot}20 \times 10^{8}$	$1{\cdot}24 \times 10^{6}$	Nuclear transitions
X-rays	10^{-10}	1 Å	10^{10}	$3{\cdot}00 \times 10^{12}$	$1{\cdot}20 \times 10^{6}$	$1{\cdot}24 \times 10^{4}$	Inner electron transitions
Ultra-violet	10^{-8}	10^{2} Å	10^{8}	$3{\cdot}00 \times 10^{10}$	$1{\cdot}20 \times 10^{4}$	$1{\cdot}24 \times 10^{2}$	Outer electron transitions
Visible	10^{-6}	1 μm	10^{6}	$3{\cdot}00 \times 10^{8}$	$1{\cdot}20 \times 10^{2}$	$1{\cdot}24$	
Infra-red							Vibration transitions
	10^{-4}	10^{2} μm	10^{4}	$3{\cdot}00 \times 10^{6}$	$1{\cdot}20$	$1{\cdot}24 \times 10^{-2}$	
							Rotational transitions
Micro-waves	10^{-2}	10^{4} μm	10^{2}	$3{\cdot}00 \times 10^{4}$	$1{\cdot}20 \times 10^{-2}$	$1{\cdot}24 \times 10^{-4}$	Electron spin transitions
Tele-vision waves	1	10^{6} μm	1	$3{\cdot}00 \times 10^{2}$	$1{\cdot}20 \times 10^{-4}$	$1{\cdot}24 \times 10^{-6}$	Nuclear spin transitions
Radio waves	10^{2}	10^{8} μm	10^{-2}	$3{\cdot}00$	$1{\cdot}20 \times 10^{-6}$	$1{\cdot}24 \times 10^{-8}$	

The Visible Spectrum

	Wavelength λ/m	Wave number σ/m^{-1}	Energy quantum $Lh\nu/\mathrm{kJ\ mol^{-1}}$	$Lh\nu/\mathrm{eV\ mol^{-1}}$
Violet	4.00×10^{-7}	$2{\cdot}50 \times 10^{6}$	299	$3{\cdot}10$
Blue				
Green	$5{\cdot}00 \times 10^{-7}$	$2{\cdot}00 \times 10^{6}$	239	$2{\cdot}48$
Yellow	$5{\cdot}89 \times 10^{-7}$ (sodium D line)			
Orange	$6{\cdot}00 \times 10^{-7}$	$1{\cdot}67 \times 10^{6}$	199	$2{\cdot}06$
Red	$7{\cdot}00 \times 10^{-7}$	$1{\cdot}43 \times 10^{6}$	171	$1{\cdot}77$

4 The Periodic Classification (with atomic numbers)

Group	I	II																		III	IV	V	VI	VII	VIII
Period																									
1	s-block								1 H	2 He														p-block	
2	3 Li	4 Be																		5 B	6 C	7 N	8 O	9 F	10 Ne
3	11 Na	12 Mg																		13 Al	14 Si	15 P	16 S	17 Cl	18 Ar
4	19 K	20 Ca	21 Sc	d-block							22 Ti	23 V	24 Cr	25 Mn	26 Fe	27 Co	28 Ni	29 Cu	30 Zn	31 Ga	32 Ge	33 As	34 Se	35 Br	36 Kr
5	37 Rb	38 Sr	39 Y								40 Zr	41 Nb	42 Mo	43 Tc	44 Ru	45 Rh	46 Pd	47 Ag	48 Cd	49 In	50 Sn	51 Sb	52 Te	53 I	54 Xe
6	55 Cs	56 Ba	57 La	58 Ce 59 Pr 60 Nd 61 Pm 62 Sm 63 Eu 64 Gd 65 Tb 66 Dy 67 Ho 68 Er 69 Tm 70 Yb 71 Lu							72 Hf	73 Ta	74 W	75 Re	76 Os	77 Ir	78 Pt	79 Au	80 Hg	81 Tl	82 Pb	83 Bi	84 Po	85 At	86 Rn
7	87 Fr	88 Ra	89 Ac	90 Th 91 Pa 92 U 93 Np 94 Pu 95 Am 96 Cm 97 Bk 98 Cf 99 Es 100 Fm 101 Md 102 No 103 Lr							104 –	105 –	106 –	107 –											

f-block: 58 Ce 59 Pr 60 Nd 61 Pm 62 Sm 63 Eu 64 Gd 65 Tb 66 Dy 67 Ho 68 Er 69 Tm 70 Yb 71 Lu

5 Electronic Configurations of the Elements (ground states)

Shell		K	L		M			N			
		$n = 1$	2		3			4			
Atomic number	Element	1s	2s	2p	3s	3p	3d	4s	4p	4d	4f
1	Hydrogen	1									
2	Helium	2									
3	Lithium	2	1								
4	Beryllium	2	2								
5	Boron	2	2	1							
6	Carbon	2	2	2							
7	Nitrogen	2	2	3							
8	Oxygen	2	2	4							
9	Fluorine	2	2	5							
10	Neon	2	2	6							
11	Sodium	2	2	6	1						
12	Magnesium	2	2	6	2						
13	Aluminium	2	2	6	2	1					
14	Silicon	2	2	6	2	2					
15	Phosphorus	2	2	6	2	3					
16	Sulphur	2	2	6	2	4					
17	Chlorine	2	2	6	2	5					
18	Argon	2	2	6	2	6					
19	Potassium	2	2	6	2	6		1			
20	Calcium	2	2	6	2	6		2			
21	Scandium	2	2	6	2	6	1	2			
22	Titanium	2	2	6	2	6	2	2			
23	Vanadium	2	2	6	2	6	3	2			
24	Chromium	2	2	6	2	6	5	1			
25	Manganese	2	2	6	2	6	5	2			
26	Iron	2	2	6	2	6	6	2			
27	Cobalt	2	2	6	2	6	7	2			
28	Nickel	2	2	6	2	6	8	2			
29	Copper	2	2	6	2	6	10	1			
30	Zinc	2	2	6	2	6	10	2			
31	Gallium	2	2	6	2	6	10	2	1		
32	Germanium	2	2	6	2	6	10	2	2		
33	Arsenic	2	2	6	2	6	10	2	3		
34	Selenium	2	2	6	2	6	10	2	4		
35	Bromine	2	2	6	2	6	10	2	5		
36	Krypton	2	2	6	2	6	10	2	6		

Atomic number	Element	Shell K $n=1$	L 2	M 3	N 4s	4p	4d	4f	O 5s	5p	5d	5f	P 6s	6p	6d
37	Rubidium	2	8	18	2	6			1						
38	Strontium	2	8	18	2	6			2						
39	Yttrium	2	8	18	2	6	1		2						
40	Zirconium	2	8	18	2	6	2		2						
41	Niobium	2	8	18	2	6	4		1						
42	Molybdenum	2	8	18	2	6	5		1						
43	Technetium	2	8	18	2	6	6		1						
44	Ruthenium	2	8	18	2	6	7		1						
45	Rhodium	2	8	18	2	6	8		1						
46	Palladium	2	8	18	2	6	10								
47	Silver	2	8	18	2	6	10		1						
48	Cadmium	2	8	18	2	6	10		2						
49	Indium	2	8	18	2	6	10		2	1					
50	Tin	2	8	18	2	6	10		2	2					
51	Antimony	2	8	18	2	6	10		2	3					
52	Tellurium	2	8	18	2	6	10		2	4					
53	Iodine	2	8	18	2	6	10		2	5					
54	Xenon	2	8	18	2	6	10		2	6					
55	Caesium	2	8	18	2	6	10		2	6			1		
56	Barium	2	8	18	2	6	10		2	6			2		
57	Lanthanum	2	8	18	2	6	10		2	6	1		2		
58	Cerium	2	8	18	2	6	10	2	2	6			2		
59	Praseodymium	2	8	18	2	6	10	3	2	6			2		
60	Neodymium	2	8	18	2	6	10	4	2	6			2		
61	Promethium	2	8	18	2	6	10	5	2	6			2		
62	Samarium	2	8	18	2	6	10	6	2	6			2		
63	Europium	2	8	18	2	6	10	7	2	6			2		
64	Gadolinium	2	8	18	2	6	10	7	2	6	1		2		
65	Terbium	2	8	18	2	6	10	9	2	6			2		
66	Dysprosium	2	8	18	2	6	10	10	2	6			2		
67	Holmium	2	8	18	2	6	10	11	2	6			2		
68	Erbium	2	8	18	2	6	10	12	2	6			2		
69	Thulium	2	8	18	2	6	10	13	2	6			2		
70	Ytterbium	2	8	18	2	6	10	14	2	6			2		
71	Lutetium	2	8	18	2	6	10	14	2	6	1		2		
72	Hafnium	2	8	18	2	6	10	14	2	6	2		2		
73	Tantalum	2	8	18	2	6	10	14	2	6	3		2		
74	Tungsten	2	8	18	2	6	10	14	2	6	4		2		
75	Rhenium	2	8	18	2	6	10	14	2	6	5		2		
76	Osmium	2	8	18	2	6	10	14	2	6	6		2		
77	Iridium	2	8	18	2	6	10	14	2	6	9				
78	Platinum	2	8	18	2	6	10	14	2	6	9		1		
79	Gold	2	8	18	2	6	10	14	2	6	10		1		
80	Mercury	2	8	18	2	6	10	14	2	6	10		2		
81	Thallium	2	8	18	2	6	10	14	2	6	10		2	1	
82	Lead	2	8	18	2	6	10	14	2	6	10		2	2	
83	Bismuth	2	8	18	2	6	10	14	2	6	10		2	3	
84	Polonium	2	8	18	2	6	10	14	2	6	10		2	4	
85	Astatine	2	8	18	2	6	10	14	2	6	10		2	5	
86	Radon	2	8	18	2	6	10	14	2	6	10		2	6	

Shell	K	L	M	N	O				P			Q
	n = 1	2	3	4	5				6			7
Atomic Element number					5s	5p	5d	5f	6s	6p	6d	7s
87 Francium	2	8	18	32	2	6	10		2	6		1
88 Radium	2	8	18	32	2	6	10		2	6		2
89 Actinium	2	8	18	32	2	6	10		2	6	1	2
90 Thorium	2	8	18	32	2	6	10		2	6	2	2
91 Protoactinium	2	8	18	32	2	6	10	2	2	6	1	2
92 Uranium	2	8	18	32	2	6	10	3	2	6	1	2
93 Neptunium	2	8	18	32	2	6	10	4	2	6	1	2
94 Plutonium	2	8	18	32	2	6	10	6	2	6		2
95 Americium	2	8	18	32	2	6	10	7	2	6		2
96 Curium	2	8	18	32	2	6	10	7	2	6	1	2
97 Berkelium	2	8	18	32	2	6	10	8	2	6	1	2
98 Californium	2	8	18	32	2	6	10	10	2	6		2
99 Einsteinium	2	8	18	32	2	6	10	11	2	6		2
100 Fermium	2	8	18	32	2	6	10	12	2	6		2
101 Mendelevium	2	8	18	32	2	6	10	13	2	6		2
102 Nobelium	2	8	18	32	2	6	10	14	2	6		2
103 Lawrencium	2	8	18	32	2	6	10	14	2	6	1	2

6 Energy Levels for a Many-Electron Atom

The following diagram gives an approximate indication of the relative energies for the various orbitals as well as the number of orbitals available in the different energy levels.

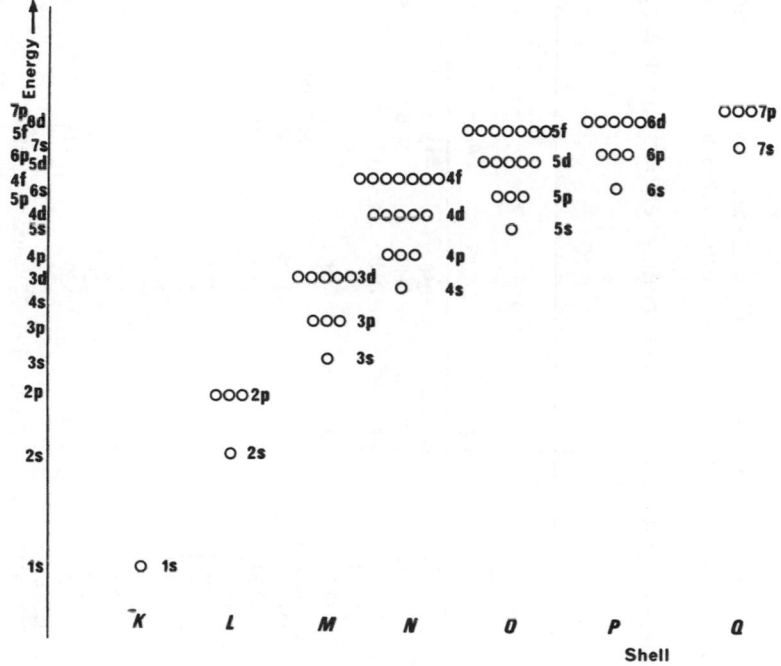

7 Properties of the Elements

The relative atomic masses in the following table are based on the $^{12}C = 12$ scale; a value in brackets denotes the mass number of the most stable isotope. The data are based on the most recent values adopted by IUPAC (1983), with a maximum of six significant figures.

ρ denotes density, $\theta_{C,m}$ denotes melting temperature, $\theta_{C,b}$ denotes boiling temperature and c_p denotes specific heat capacity.

subl. denotes sublimes

Element	Symbol	Atomic number	Relative atomic mass	ρ/g cm^{-3}	$\theta_{C,m}$/°C	$\theta_{C,b}$/°C	c_p/J kg^{-1} K^{-1}	Oxidation states
Actinium	Ac	89	227·028	10·1	1050	3200		3
Aluminium	Al	13	26·9815	2·70	660	2470	900	3
Americium	Am	95	(243)	11·7	(1200)	(2600)	140	3,4,5,6
Antimony	Sb	51	121·75	6·62	630	1380	209	3,5
Argon	Ar	18	39·948	1·40 (87 K)	−189	−186	519	
Arsenic (α, grey)	As	33	74·9216	5·72		613 subl.	326	3,5
Astatine	At	85	(210)		(302)	(380)	(140)	
Barium	Ba	56	137·33	3·51	714	1640	192	2
Berkelium	Bk	97	(247)					3,4
Beryllium	Be	4	9·01218	1·85	1280	2477	$1·82 \times 10^3$	2
Bismuth	Bi	83	208·980	9·80	271	1560	121	3,5
Boron	B	5	10·81	2·34	2300	3930	$1·03 \times 10^3$	3
Bromine	Br	35	79·904	3·12	−7·2	58·8	448	1,3,4,5,6
Cadmium	Cd	48	112·41	8·64	321	765	230	2
Caesium	Cs	55	132·905	1·90	28·7	690	234	1
Calcium	Ca	20	40·08	1·54	850	1487	653	2
Californium	Cf	98	(251)					3
Carbon (graphite)	C	6	12·011	2·25 (graphite) 3·51 (diamond)	3730 subl.	4830	711 (graphite) 519 (diamond)	2,4

Element	Symbol	Atomic number	Relative atomic mass	$\rho/\text{g cm}^{-3}$	$\theta_{C,m}/^{\circ}C$	$\theta_{C,b}/^{\circ}C$	$c_p/\text{J kg}^{-1}\text{ K}^{-1}$	Oxidation states
Cerium	Ce	58	140·12	6·78	795	3470	184	3,4
Chlorine	Cl	17	35·453	1·56 (238 K)	−101	−34·7	477	1,3,4,5,6,7
Chromium	Cr	24	51·996	7·19	1890	2482	448	2,3,6
Cobalt	Co	27	58·9332	8·90	1492	2900	435	2,3
Copper	Cu	29	63·546	8·92	1083	2595	385	1,2
Curium	Cm	96	(247)					3
Dysprosium	Dy	66	162·50	8·56	1410	2600	172	3
Einsteinium	Es	99	(252)					3
Erbium	Er	68	167·26	9·16	1500	2900	167	3
Europium	Eu	63	151·96	5·24	826	1440	138	2,3
Fermium	Fm	100	(257)					3
Fluorine	F	9	18·9984	1·11 (85 K)	−220	−188	824	1
Francium	Fr	87	(223)		(27)	(680)	(140)	1
Gadolinium	Gd	64	157·25	7·95	1310	3000	234	3
Gallium	Ga	31	69·72	5·91	29·8	2400	381	3
Germanium	Ge	32	72·59	5·35	937	2830	322	4
Gold	Au	79	196·967	19·3	1063	2970	130	1,3
Hafnium	Hf	72	178·49	13·3	2220	5400	146	4
Helium	He	2	4·0026	0·147 (4 K)	−270	−269	$5·19 \times 10^3$	
Holmium	Ho	67	164·930	8·80	1460	2600	163	3
Hydrogen	H	1	1·0079	0·070 (20 K)	−259	−252	$1·43 \times 10^4$	1
Indium	In	49	114·82	7·30	157	2000	238	1,3
Iodine	I	53	126·905	4·93	114	184	218	1,3,5,7
Iridium	Ir	77	192·22	22·5	2440	5300	134	2,3,4,6
Iron	Fe	26	55·847	7·86	1535	3000	448	2,3,6
Krypton	Kr	36	83·80	2·16 (121 K)	−157	−152	247	2
Lanthanum	La	57	138·906	6·19	920	3470	201	3
Lawrencium	Lr	103	(260)					
Lead	Pb	82	207.2	11·3	327	1744	130	2,4
Lithium	Li	3	6·941	0·53	180	1330	$3·39 \times 10^3$	1

Element	Symbol	Atomic number	Relative atomic mass	ρ/g cm^{-3}	$\theta_{c,m}$/°C	$\theta_{c,b}$/°C	c_p/J kg^{-1} K^{-1}	Oxidation states
Lutetium	Lu	71	174·967	9·84	1650	3330	155	3
Magnesium	Mg	12	24·305	1·74	650	1110	$1·03 \times 10^3$	2
Manganese	Mn	25	54·9380	7·20	1240	2100	477	2,3,4,6,7
Mendelevium	Md	101	(258)					3
Mercury	Hg	80	200·59	13·6	−38·9	357	138	1,2
Molybdenum	Mo	42	95·94	10·2	2610	5560	251	2,3,4,5,6
Neodymium	Nd	60	144·24	7·00	1020	3030	188	3
Neon	Ne	10	20·179	1·20 (27 K)	−249	−246	$1·03 \times 10^3$	
Neptunium	Np	93	237·048	20·4	640			3,4,5,6
Nickel	Ni	28	58·69	8·90	1453	2730	439	2,3
Niobium	Nb	41	92·9064	8·57	2470	3300	264	3,5
Nitrogen	N	7	14·0067	0·808 (77 K)	−210	−196	$1·04 \times 10^3$	1,2,3,4,5
Nobelium	No	102	(259)					
Osmium	Os	76	190·2	22·5	3000	5000	130	2,3,4,6,8
Oxygen	O	8	15·9994	1·15 (90 K)	−218	−183	916	2
Palladium	Pd	46	106·42	12·0	1550	3980	243	2,4
Phosphorus	P	15	30·9738	1·82 (white) 2·34 (red)	44·2 (white) 590 (red)	280 (white)	757 (white) 670 (red)	3,5
Platinum	Pt	78	195·08	21·4	1769	4530	134	2,4,6
Plutonium	Pu	94	(244)	19·8	640	3240		3,4,5,6
Polonium	Po	84	(209)	9·4	254	960	126	2,4
Potassium	K	19	39·0983	0·86	63·7	774	753	1
Praseodymium	Pr	59	140·908	6·78	935	3130	192	3,4
Promethium	Pm	61	(145)		1030	2730	184	3
Protoactinium	Pa	91	231·036	15·4	1230		121	4,5
Radium	Ra	88	226·025	5·0	700	1140	121	2
Radon	Rn	86	(222)	4·4 (211 K)	−71	−61·8	92	
Rhenium	Re	75	186·207	20·5	3180	5630	138	2,4,5,6,7
Rhodium	Rh	45	102·906	12·4	1970	4500	243	2,3,4
Rubidium	Rb	37	85·4678	1·53	38·9	688	360	1

Element	Symbol	Atomic number	Relative atomic mass	ρ/g cm^{-3}	$\theta_{C,m}$/°C	$\theta_{C,b}$/°C	c_p/J kg^{-1} K^{-1}	Oxidation states
Ruthenium	Ru	44	101·07	12·3	2500	4900	238	3,4,5,6,8
Samarium	Sm	62	150·36	7·54	1070	1900	197	2,3
Scandium	Sc	21	44·9559	2·99	1540	2730	556	3
Selenium	Se	34	78·96	4·81	217	685	322	2,4,6
Silicon	Si	14	28·0855	2·33	1410	2360	711	4
Silver	Ag	47	107·868	10·5	961	2210	234	1
Sodium	Na	11	22·9898	0·97	97·8	890	$1·23 \times 10^3$	1
Strontium	Sr	38	87·62	2·62	768	1380	284	2
Sulphur (α, rhombic)	S	16	32·06	2·07 (α) / 1·96 (β)	113 (α) / 119 (β)	445	732	2,4,6
Tantalum	Ta	73	180·948	16·6	3000	5420	138	5
Technetium	Tc	43	(98)	11·5	2200	3500	243	7
Tellurium	Te	52	127·60	6·25	450	990	201	2,4,6
Terbium	Tb	65	158·925	8·27	1360	2800	184	3,4
Thallium	Tl	81	204·383	11·8	304	1460	130	1,3
Thorium	Th	90	232·038	11·7	1750	3850	113	3,4
Thulium	Tm	69	168·934	9·33	1540	1730	159	2,3
Tin (white)	Sn	50	118·71	7·28 (white) / 5·75 (grey)	232	2270	218	2,4
Titanium	Ti	22	47·88	4·54	1675	3260	523	2,3,4
Tungsten	W	74	183·85	19·4	3410	5930	134	2,4,5,6
Uranium	U	92	238·029	19·1	1130	3820	117	3,4,5,6
Vanadium	V	23	50·9415	5·96	1900	3000	481	2,3,4,5
Xenon	Xe	54	131·29	3·52 (165 K)	−112	−108	159	2,4,6,8
Ytterbium	Yb	70	173·04	6·98	824	1430	146	2,3
Yttrium	Y	39	88·9059	4·34	1500	2930	297	3
Zinc	Zn	30	65·39	7·14	420	907	385	2
Zirconium	Zr	40	91·224	6·49	1850	3580	276	2,3,4

8 Crystal Structures of the Solid Elements

b denotes body-centred cubic   *h* denotes hexagonal close-packed
f denotes face-centred cubic   *x* denotes complex

							H	He									
							h	*h*									
Li	Be											B	C	N	O	F	Ne
b	*h*											*x*					*f*
Na	Mg											Al	Si	P	S	Cl	Ar
b	*h*											*f*					*f*
K	Ca	Sc	Ti	V	Cr	Mn	Fe	Co	Ni	Cu	Zn	Ga	Ge	As	Se	Br	Kr
b	*f*	*f*	*h*	*b*	*b*	*x*	*b*	*h*	*f*	*f*	*h*	*x*	*x*	*x*	*x*		*f*
Rb	Sr	Y	Zr	Nb	Mo	Tc	Ru	Rh	Pd	Ag	Cd	In	Sn	Sb	Te	I	Xe
b	*f*	*b*	*h*	*b*	*b*	*h*	*h*	*f*	*f*	*f*	*h*	*x*	*x*	*x*	*x*		*f*
Cs	Ba	La	Hf	Ta	W	Re	Os	Ir	Pt	Au	Hg	Tl	Pb	Bi	Po	At	Rn
b	*b*	*f*	*h*	*b*	*b*	*h*	*h*	*f*	*f*	*f*	*x*	*h*	*f*	*x*	*x*		*f*
Fr	Ra	Ac															
		f															

	Ce	Pr	Nd	Pm	Sm	Eu	Gd	Tb	Dy	Ho	Er	Tm	Yb	Lu
	f	*f*	*h*	*h*	*h*	*b*	*h*	*h*	*h*	*h*	*h*	*h*	*f*	*h*
	Th	Pa	U											
	f	*x*	*x*											

9 Stable Isotopes of the Elements

* Indicates a radioactive isotope with a half-life $> 10^4$ years.

Atomic number	Isotope	Abundance (%)	Atomic number	Isotope	Abundance (%)
1	^{1}H	99·985		^{48}Ca*	0·18
	^{2}H	0·015	21	^{45}Sc	100
2	^{3}He	0·00013	22	^{46}Ti	8·0
	^{4}He	ca.100		^{47}Ti	7·3
3	^{6}Li	7·4		^{48}Ti	73·8
	^{7}Li	92·6		^{49}Ti	5·5
4	^{9}Be	100		^{50}Ti	5·3
5	^{10}B	18·7	23	^{50}V*	0·25
	^{11}B	81·3		^{51}V	99 8
6	^{12}C	98·9	24	^{50}Cr	4·3
	^{13}C	1·1		^{52}Cr	83·8
7	^{14}N	99·6		^{53}Cr	9·5
	^{15}N	0·37		^{54}Cr	2·4
8	^{16}O	99·8	25	^{55}Mn	100
	^{17}O	0·037	26	^{54}Fe	5·8
	^{18}O	0·20		^{56}Fe	91·6
9	^{19}F	100		^{57}Fe	2·2
10	^{20}Ne	90·9		^{58}Fe	0·33
	^{21}Ne	0·26	27	^{59}Co	100
	^{22}Ne	8·8	28	^{58}Ni	67·9
11	^{23}Na	100		^{60}Ni	26·2
12	^{24}Mg	78 6		^{61}Ni	1·2
	^{25}Mg	10·1		^{62}Ni	3·7
	^{26}Mg	11·3		^{64}Ni	1·0
13	^{27}Al	100	29	^{63}Cu	69·1
14	^{28}Si	92·2		^{65}Cu	30·9
	^{29}Si	4·7	30	^{64}Zn	48·9
	^{30}Si	3·1		^{66}Zn	27·8
15	^{31}P	100		^{67}Zn	4·1
16	^{32}S	95·0		^{68}Zn	18·6
	^{33}S	0·76		^{70}Zn	0·62
	^{34}S	4·2	31	^{69}Ga	60·4
	^{36}S	0·021		^{71}Ga	39·6
17	^{35}Cl	75·5	32	^{70}Ge	20·6
	^{37}Cl	24·5		^{72}Ge	27·4
18	^{36}Ar	0·34		^{73}Ge	7·8
	^{38}Ar	0·063		^{74}Ge	36·5
	^{40}Ar	99·6		^{76}Ge	7·7
19	^{39}K	93·1	33	^{75}As	100
	^{40}K*	0·012	34	^{74}Se	0·87
	^{41}K	6·9		^{76}Se	9·0
20	^{40}Ca	96·9		^{77}Se	7·6
	^{42}Ca	0·64		^{78}Se	23·5
	^{43}Ca	0·14		^{80}Se	49·8
	^{44}Ca	2·1		^{82}Se	9·2
	^{46}Ca	0·0032	35	^{79}Br	50·5

Atomic number	Isotope	Abundance (%)	Atomic number	Isotope	Abundance (%)
	^{81}Br	49·5	49	^{113}In	4·2
36	^{78}Kr	0·35		^{115}In*	95·8
	^{80}Kr	2·3	50	^{112}Sn	1·0
	^{82}Kr	11·6		^{114}Sn	0·65
	^{83}Kr	11·5		^{115}Sn	0·35
	^{84}Kr	56·9		^{116}Sn	14·2
	^{86}Kr	17·4		^{117}Sn	7·6
37	^{85}Rb	72·2		^{118}Sn	24·0
	^{87}Rb*	27·8		^{119}Sn	8·6
38	^{84}Sr	0·56		^{120}Sn	32·8
	^{86}Sr	9·9		^{122}Sn	4·8
	^{87}Sr	7·0		^{124}Sn	6·0
	^{88}Sr	82·6	51	^{121}Sb	57·2
39	^{89}Y	100		^{123}Sb	42·8
40	^{90}Zr	51·5	52	^{120}Te	0·089
	^{91}Zr	11·2		^{122}Te	2·5
	^{92}Zr	17·1		^{123}Te	0·89
	^{94}Zr	17·4		^{124}Te	4·6
	^{96}Zr	2·8		^{125}Te	7·0
41	^{93}Nb	100		^{126}Te	18·7
42	^{92}Mo	15·1		^{128}Te	31·8
	^{94}Mo	9·3		^{130}Te	34·5
	^{95}Mo	15·8	53	^{127}I	100
	^{96}Mo	16·5	54	^{124}Xe	0·095
	^{97}Mo	9·6		^{126}Xe	0·088
	^{98}Mo	24·0		^{128}Xe	1·9
	^{100}Mo	9·7		^{129}Xe	26·2
43	(Tc)			^{130}Xe	4·1
44	^{96}Ru	5·5		^{131}Xe	21·2
	^{98}Ru	1·9		^{132}Xe	26·9
	^{99}Ru	12·7		^{134}Xe	10·5
	^{100}Ru	12·7		^{136}Xe	8·9
	^{101}Ru	17·0	55	^{133}Cs	100
	^{102}Ru	31·6	56	^{130}Ba	0·10
	^{104}Ru	18·6		^{132}Ba	0·097
45	^{103}Rh	100		^{134}Ba	2·4
46	^{102}Pd	0·96		^{135}Ba	6·6
	^{104}Pd	11·0		^{136}Ba	7·8
	^{105}Pd	22·2		^{137}Ba	11·3
	^{106}Pd	27·3		^{138}Ba	71·7
	^{108}Pd	26·7	57	^{138}La*	0·089
	^{110}Pd	11·8		^{139}La	99·9
47	^{107}Ag	51·4	58	^{136}Ce	0·19
	^{109}Ag	48·6		^{138}Ce	0·25
48	^{106}Cd	1·2		^{140}Ce	88·5
	^{108}Cd	0·89		^{142}Ce	11·1
	^{110}Cd	12·4	59	^{141}Pr	100
	^{111}Cd	12·8	60	^{142}Nd	27·1
	^{112}Cd	24·1		^{143}Nd	12·2
	^{113}Cd	12·3		^{144}Nd*	23·9
	^{114}Cd	28·8		^{145}Nd	8·3
	^{116}Cd	7·6		^{146}Nd	17·2

Atomic number	Isotope	Abundance (%)	Atomic number	Isotope	Abundance (%)
	^{148}Nd	5·7	73	^{180}Ta*	0·012
	^{150}Nd	5·6		^{181}Ta	99·99
61	(Pm)		74	^{180}W	0·13
62	^{144}Sm	3·1		^{182}W	26·3
	^{147}Sm*	15·1		^{183}W	14·3
	^{148}Sm	11·3		^{184}W	30·7
	^{149}Sm	13·9		^{186}W	28·6
	^{150}Sm	7·5	75	^{185}Re	37·1
	^{152}Sm	26·6		^{187}Re*	62·9
	^{154}Sm	22·5	76	^{184}Os	0·018
63	^{151}Eu	47·8		^{186}Os	1·6
	^{153}Eu	52·2		^{187}Os	1·6
64	^{152}Gd	0·20		^{188}Os	13·3
	^{154}Gd	2·2		^{189}Os	16·1
	^{155}Gd	14·8		^{190}Os	26·4
	^{156}Gd	20·6		^{192}Os	41·0
	^{157}Gd	15·7	77	^{191}Ir	38·5
	^{158}Gd	24·8		^{193}Ir	61·5
	^{160}Gd	21·7	78	^{190}Pt*	0·012
65	^{159}Tb	100		^{192}Pt	0·78
66	^{156}Dy	0·052		^{194}Pt	32·8
	^{158}Dy	0·090		^{195}Pt	33·7
	^{160}Dy	2·3		^{196}Pt	25·4
	^{161}Dy	18·9		^{198}Pt	7·2
	^{162}Dy	25·5	79	^{197}Au	100
	^{163}Dy	25·0	80	^{196}Hg	0·15
	^{164}Dy	28·2		^{198}Hg	10·0
67	^{165}Ho	100		^{199}Hg	16·8
68	^{162}Er	0·14		^{200}Hg	23·1
	^{164}Er	1·5		^{201}Hg	13·2
	^{166}Er	33·4		^{202}Hg	29·8
	^{167}Er	22·9		^{204}Hg	6·8
	^{168}Er	27·1	81	^{203}Tl	29·5
	^{170}Er	14·9		^{205}Tl	70·5
69	^{169}Tm	100	82	^{204}Pb	1·5
70	^{168}Yb	0·14		^{206}Pb	23·6
	^{170}Yb	3·0		^{207}Pb	22·6
	^{171}Yb	14·3		^{208}Pb	52·3
	^{172}Yb	21·9	83	^{209}Bi	100
	^{173}Yb	16·2	84	(Po)	
	^{174}Yb	31·8	85	(At)	
	^{176}Yb	12·6	86	(Rn)	
71	^{175}Lu	97·4	87	(Fr)	
	^{176}Lu*	2·6	88	(Ra)	
72	^{174}Hf	0·18	89	(Ac)	
	^{176}Hf	5·2	90	^{232}Th	100
	^{177}Hf	18·5	91	(Pa)	
	^{178}Hf	27·1	92	^{234}U*	0·0056
	^{179}Hf	13·8		^{235}U*	0·72
	^{180}Hf	35·2		^{238}U*	99·3

10 Half-lives of Radioactive Elements

The following symbols are used: s for second, min for minute, h for hour, d for day and a for year.

$1 \text{ min} = 60 \text{ s}$ $1 \text{ h} = 3{\cdot}6 \text{ ks}$ $1 \text{ d} = 86{\cdot}4 \text{ ks}$ $1 \text{ a} \approx 31{\cdot}6 \text{ Ms}$

Isotope	Particle emitted	Half-life	Isotope	Particle emitted	Half-life
^{3}H	β^{-}	12·3 a	^{210}At	α	8·3 h
^{14}C	β^{-}	5570 a	^{242}Pu	α	$3{\cdot}79 \times 10^{5}$ a
^{24}Na	β^{-}	15·0 h	^{243}Am	α	$7{\cdot}95 \times 10^{3}$ a
^{32}P	β^{-}	14·3 d	^{247}Cm	α	$1{\cdot}64 \times 10^{7}$ a
^{35}S	β^{-}	86·7 d	^{247}Bk	α	7×10^{3} a
^{36}Cl	β^{-}	3×10^{5} a	^{251}Cf	α	800 a
^{40}K	β^{-}	$1{\cdot}3 \times 10^{9}$ a	^{254}Es	α	280 d
^{60}Co	β^{-}	5·23 a	^{253}Fm	α	4·5 d
^{90}Sr	β^{-}	27 a	^{256}Md	spontaneous fission	1·5 h
^{99}Tc	β^{-}	$2{\cdot}2 \times 10^{5}$ a			
^{131}I	β^{-}	8·06 d	^{253}No	α	10 min
^{147}Pm	β^{-}	2·64 a	^{257}Lr	α	8 s

The particle emitted is denoted as follows: α, α-particle; β⁻, β-particle. Half-lives are shown in parentheses.
The following symbols are used: s for second, min for minute (1 min = 60 s), h for hour (1 h = 3·6 ks), d for day (1 d = 86·4 ks) and a for year (1 a ≈ 31·6 Ms).

Thorium series — *Neptunium series* — *Uranium series* — *Actinium series*

Thorium series

$^{232}_{90}$Th → α (1·39 × 10¹⁰ a)
$^{228}_{88}$Ra → β⁻ (6·7 a)
$^{228}_{89}$Ac → β⁻ (6·13 h)
$^{228}_{90}$Th → α (1·91 a)
$^{224}_{88}$Ra → α (3·64 d)
$^{220}_{86}$Rn → α (51·5 s)
$^{216}_{84}$Po
  α (0·158 s) / β⁻
$^{212}_{82}$Pb (10·6 h) β⁻ — $^{216}_{85}$At α (3 × 10⁻⁴ s)
$^{212}_{83}$Bi
  α (60·5 min) / β⁻
$^{208}_{81}$Tl (3·10 min) β⁻ — $^{212}_{84}$Po α (3 × 10⁻⁷ s)
$^{208}_{82}$Pb

Neptunium series

$^{237}_{93}$Np → α (2·20 × 10⁶ a)
$^{233}_{91}$Pa → β⁻ (27·4 d)
$^{233}_{92}$U → α (1·62 × 10⁵ a)
$^{229}_{90}$Th → α (7340 a)
$^{225}_{88}$Ra → β⁻ (14·8 d)
$^{225}_{89}$Ac → α (10·0 d)
$^{221}_{87}$Fr → α (4·8 min)
$^{217}_{85}$At → α (0·018 s)
$^{213}_{83}$Bi
  α (47 min) / β⁻
$^{209}_{81}$Tl (2·2 min) β⁻ — $^{213}_{84}$Po α (4·2 × 10⁻⁶ s)
$^{209}_{82}$Pb → β⁻ (3·3 h)
$^{209}_{83}$Bi

Uranium series

$^{238}_{92}$U → α (4·51 × 10⁹ a)
$^{234}_{90}$Th → β⁻ (24·1 d)
$^{234}_{91}$Pa → β⁻ (1·17 min)
$^{234}_{92}$U → α (2·52 × 10⁵ a)
$^{230}_{90}$Th → α (8·0 × 10⁴ a)
$^{226}_{88}$Ra → α (1622 a)
$^{222}_{86}$Rn → α (3·825 d)
$^{218}_{84}$Po
  α (3·05 min) / β⁻
$^{214}_{82}$Pb (26·8 min) β⁻ — $^{218}_{85}$At α (1·35 s)
$^{214}_{83}$Bi
  α (19·7 min) / β⁻
$^{210}_{81}$Tl (1·3 min) β⁻ — $^{214}_{84}$Po α (1·64 × 10⁻⁴ s)
$^{210}_{82}$Pb → β⁻ (21 a)
$^{210}_{83}$Bi
  α (5·0 d) / β⁻
$^{206}_{81}$Tl (4·20 min) β⁻ — $^{210}_{84}$Po α (138·4 d)
$^{206}_{82}$Pb

Actinium series

$^{235}_{92}$U → α (7·13 × 10⁸ a)
$^{231}_{90}$Th → β⁻ (24·6 h)
$^{231}_{91}$Pa → α (3·43 × 10⁴ a)
$^{227}_{89}$Ac
  α (21·6 a) / β⁻
$^{223}_{87}$Fr (22 min) β⁻ — $^{227}_{90}$Th α (18·2 d)
$^{223}_{88}$Ra → α (11·7 d)
$^{219}_{86}$Rn → α (3·92 s)
$^{215}_{84}$Po
  α (1·83 × 10⁻³ s) / β⁻
$^{211}_{82}$Pb (36·1 min) β⁻ — $^{215}_{85}$At α (10⁻⁴ s)
$^{211}_{83}$Bi
  α (2·16 min) / β⁻
$^{207}_{81}$Tl (4·8 min) β⁻ — $^{211}_{84}$Po α (0·52 s)
$^{207}_{82}$Pb

12 Electric Conductivities of the Elements at 298 K

The following table gives values of $10^{-8} \kappa / \mathrm{S \ m^{-1}}$ where κ denotes electric conductivity and siemens $\mathrm{S} = \Omega^{-1}$.

H	He

Li	Be											B	C[a]	N	O	F
0·108	0·25											10^{-12}	0·0007			
Na	Mg											Al	Si	P	S	Cl
0·218	0·224											0·382	$2 \times 10^{-10} -10^{-7b}$	10^{-17}	10^{-23}	
K	Ca	Sc	Ti	V	Cr	Mn	Fe	Co	Ni	Cu	Zn	Ga	Ge	As	Se	Br
0·143	0·218	0·015	0·024	0·04	0·078	0·054	0·10	0·16	0·145	0·593	0·167	0·058	$2·2 \times 10^{-8}$	0·029	0·08	10^{-18}
Rb	Sr	Y	Zr	Nb	Mo	Tc	Ru	Rh	Pd	Ag	Cd	In	Sn	Sb	Te	I
0·080	0·043	0·018	0·024	0·080	0·19	10^{-5}	0·10	0·22	0·093	0·62	0·146	0·111	0·088	0·026	10^{-6}	10^{-15}
Cs	Ba	La	Hf	Ta	W	Re	Os	Ir	Pt	Au	Hg	Tl	Pb	Bi	Po	At
0·053	0·016	0·017	0·028	0·081	0·18	0·051	0·11	0·19	0·094	0·42	0·010	0·055	0·046	0·009	0·02	
Fr	Ra	Ac														

Ce	Pr	Nd	Pm	Sm	Eu	Gd	Tb	Dy	Ho	Er	Tm	Yb	Lu
0·013	0·015	0·013		0·011	0·011	0·007	0·009	0·011	0·011	0·012	0·011	0·034	0·013
Th	Pa	U											
0·055		0·033											

[a] Graphite.   [b] Considerable discrepancy between sources.

13 Ionization Energies of the Elements

Note. The values of $\Delta U/\text{kJ mol}^{-1}$ in the following tables strictly relate to 0 K and conversion to $\Delta H/\text{kJ mol}^{-1}$ at 298 K requires the addition of $(5RT/2)\text{ kJ mol}^{-1}$, that is approximately 6. Thermodynamic energy U is related to enthalpy H by

$$H = U + pV.$$

(a) First Ionization Energies

The minimum energy required to remove an electron from an isolated atom in the gaseous state is known as the first ionization energy of the element, i.e. it is the increase in energy for the process:

$$M(g) = M^+(g) + e^-$$

H	He
1310	2370

Li	Be											B	C	N	O	F	Ne
519	900											799	1090	1400	1310	1680	2080

Na	Mg											Al	Si	P	S	Cl	Ar
494	736											577	786	1060	1000	1260	1520

K	Ca	Sc	Ti	V	Cr	Mn	Fe	Co	Ni	Cu	Zn	Ga	Ge	As	Se	Br	Kr
418	590	632	661	648	653	716	762	757	736	745	908	577	762	966	941	1140	1350

Rb	Sr	Y	Zr	Nb	Mo	Tc	Ru	Rh	Pd	Ag	Cd	In	Sn	Sb	Te	I	Xe
402	548	636	669	653	694	699	724	745	803	732	866	556	707	833	870	1010	1170

Cs	Ba	La	Hf	Ta	W	Re	Os	Ir	Pt	Au	Hg	Tl	Pb	Bi	Po	At	Rn
376	502	540	531	760	770	762	841	887	866	891	1010	590	716	703	812	920	1040

Fr	Ra	Ac
381	510	669

Ce	Pr	Nd	Pm	Sm	Eu	Gd	Tb	Dy	Ho	Er	Tm	Yb	Lu
665	556	607	556	540	548	594	648	657				598	481

Th	Pa	U	Np	Pu	Am	Cm	Bk	Cf	Es	Fm	Md	No	Lr
674		385											

(b) Successive Ionization Energies

The minimum energy required to remove a second electron from a unipositive ion in the gaseous state is known as the second ionization energy, i.e. it is the increase in energy for the process:

$$M^+(g) = M^{2+}(g) + e^-$$

Higher ionization energies are defined in a similar way.

Atomic number	Element	\u0394U/kJ mol^{-1}							
		1st	2nd	3rd	4th	5th	6th	7th	8th
1	H	1310							
2	He	2370	5250						
3	Li	519	7300	11800					
4	Be	900	1760	14800	21000				
5	B	799	2420	3660	25000	32800			
6	C	1090	2350	4610	6220	37800	47000		
7	N	1400	2860	4590	7480	9440	53200	64300	
8	O	1310	3390	5320	7450	11000	13300	71000	84100
9	F	1680	3370	6040	8410	11000	15100	17900	91600
10	Ne	2080	3950	6150	9290	12100	15200	20000	23000
11	Na	494	4560	6940	9540	13400	16600	20100	25500
12	Mg	736	1450	7740	10500	13600	18000	21700	25600
13	Al	577	1820	2740	11600	14800	18400	23400	27500
14	Si	786	1580	3230	4360	16000	20000	23600	29100
15	P	1060	1900	2920	4960	6280	21200	25900	30500
16	S	1000	2260	3390	4540	6990	8490	27100	31700
17	Cl	1260	2300	3850	5150	6540	9330	11000	33600
18	Ar	1520	2660	3950	5770	7240	8790	12000	13800
19	K	418	3070	4600	5860	7990	9620	11400	14900
20	Ca	590	1150	4940	6480	8120	10700	12300	14600
21	Sc	632	1240	2390	7110	8870	10700	13600	15300
22	Ti	661	1310	2720	4170	9620	11600	13600	17000
23	V	648	1370	2870	4600	6280	12400	14600	16700
24	Cr	653	1590	2990	4770	7070	8700	16600	17700
25	Mn	716	1510	3250	5190	7360	9750	11500	18800
26	Fe	762	1560	2960	5400	7620	10100	12800	14600
27	Co	757	1640	3230	5100	7910	10500	13300	16400
28	Ni	736	1750	3390	5400	7620	10900	13800	17000
29	Cu	745	1960	3550	5690	7990	10500	14300	17500
30	Zn	908	1730	3828	5980	8280	11000	13900	18100
31	Ga	577	1980	2960	6190	8700	11400	14400	17700
32	Ge	762	1540	3300	4390	8950	11900	14900	18200
33	As	966	1950	2730	4850	6020	12300	15400	18900
34	Se	941	2080	3090	4140	7030	7870	16000	19500
35	Br	1140	2080	3460	4850	5770	8370	10000	20300
36	Kr	1350	2370	3560	5020	6370	7570	10700	12200
37	Rb	402	2650	3850	5110	6850	8300	9800	12900
38	Sr	548	1060	4120	5440	6940	9000	10500	12200
39	Y	636	1180	1980	6000	7400	9100	11300	13000
40	Zr	669	1270	2220	3310	8000	9600	11000	14000
41	Nb	653	1380	2430	3690	4850	10000	12000	14000
42	Mo	694	1560	2620	4480	5400	7100	12000	15000
43	Tc	699	1470	2800	4100	5900	7500	9200	15000
44	Ru	724	1620	2740	4500	6300	7900	9600	11000
45	Rh	745	1740	3000	4400	6300	8400	10000	12000
46	Pd	803	1870	3180	4730	6300	8800	10000	13000
47	Ag	732	2070	3360	5000	6700	8400	11000	13000
48	Cd	866	1630	3620	5300	7000	9100	11100	14100
49	In	556	1820	2700	5230	7400	9500	11700	13900
50	Sn	707	1410	2940	3930	7780	9900	12200	14600
51	Sb	833	1590	2440	4270	5360	10400	12700	15200
52	Te	870	1800	3010	3680	5860	7000	13200	15800

Atomic number	Element	1st	2nd	3rd	ΔU/kJ mol^{-1} 4th	5th	6th	7th	8th
53	I	1010	1840	3000	4030	5000	7400	8700	16400
54	Xe	1170	2050	3100	4300	5800	8000	9800	12200
55	Cs	376	2420	3300	4400	6000	7100	8300	11300
56	Ba	502	966	3390	4700	6000	7700	9000	10200
57	La	540	1100	1850	5000	6400	7700	9600	11000
72	Hf	531	1440	2010	3010				
73	Ta	760	1560	2150	3190	4350			
74	W	770	1710	2330	3420	4600	5900		
75	Re	762	1600	2500	3600	5000	6300	7500	
76	Os	841	1630	2400	3800	5000	6700	8000	9600
77	Ir	887	1550	2600	3800	5400	7100	8400	10000
78	Pt	866	1870	2750	3970	5400	7200	8800	10500
79	Au	891	1980	2940	4200	5400	7100	9200	11000
80	Hg	1010	1810	3300	7000	7900			
81	Tl	590	1970	2870	4900	6200	7800	9500	11300
82	Pb	716	1450	3080	4080	6700	8100	9900	11800
83	Bi	703	1610	2460	4350	5400	8500	10300	12300
84	Po	812							
85	At	920							
86	Rn	1040	1930	2890	4250	5310			
87	Fr	381							
88	Ra	510	979						
89	Ac	669	1170						
90	Th	674	1110	1930	2760				
91	Pa								
92	U	385							

14 Electron Affinities of the Elements

The electron affinity of an element is the energy released when an electron is added to an isolated atom in the gaseous state, i.e. it is the decrease in energy for the process:

$$X(g) + e^- = X^-(g)$$

Similarly, the second electron affinity is the *increase* in energy for the process:

$$X^-(g) + e^- = X^{2-}(g)$$

Note. The following values of $\Delta U/\text{kJ mol}^{-1}$ strictly relate to 0 K and conversion to $\Delta H/\text{kJ mol}^{-1}$ at 298 K requires the addition of $(5RT/2)$ kJ mol^{-1}, that is approximately 6.

	H −72						
Li −52		B −29	C −120	N −3	O −142	O⁻ +844	F −348
Na −71			Si −180	P −70	S −200	S⁻ +532	Cl −364
							Br −342
							I −314

15 Atomic Spectrum of Hydrogen

The frequency v of any line in the atomic spectrum of hydrogen is given by

$$v = R_H c(1/n_2^2 - 1/n_1^2)$$

where R_H denotes the Rydberg constant for hydrogen and n_1 and n_2 are integers.

(a) Balmer Series

The series of lines for which $n_2 = 2$ occurs in the visible region and is known as the Balmer series.

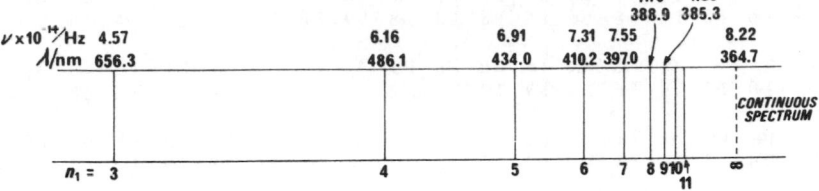

(b) Atomic Spectrum of Hydrogen

The complete atomic hydrogen spectrum also includes the Lyman ($n_2 = 1$), Paschen ($n_2 = 3$), Brackett ($n_2 = 4$) and Pfund ($n_2 = 5$) series. The frequencies of the lines in the Lyman series (ultraviolet region) are given in the following table.

n_1	$v \times 10^{-15}$/Hz
2	2·467
3	2·924
4	3·084
5	3·158
6	3·198
7	3·223
8	3·238
9	3·249
10	3·257
∞	3·290

16 Electronegativities of the Elements (Pauling)

Electronegativity is a measure of the relative tendency of an atom to attract a bonding pair of electrons. Pauling assigned values to the elements on an arbitrary scale from 0 to 4.

						H 2·1	He										
Li 1·0	Be 1·5											B 2·0	C 2·5	N 3·0	O 3·5	F 4·0	Ne
Na 0·9	Mg 1·2											Al 1·5	Si 1·8	P 2·1	S 2·5	Cl 3·0	Ar
K 0·8	Ca 1·0	Sc 1·3	Ti 1·5	V 1·6	Cr 1·6	Mn 1·5	Fe 1·8	Co 1·8	Ni 1·8	Cu 1·9	Zn 1·6	Ga 1·6	Ge 1·8	As 2·0	Se 2·4	Br 2·8	Kr
Rb 0·8	Sr 1·0	Y 1·2	Zr 1·4	Nb 1·6	Mo 1·8	Tc 1·9	Ru 2·2	Rh 2·2	Pd 2·2	Ag 1·9	Cd 1·7	In 1·7	Sn 1·8	Sb 1·9	Te 2·1	I 2·5	Xe
Cs 0·7	Ba 0·9	La 1·1	Hf 1·3	Ta 1·5	W 1·7	Re 1·9	Os 2·2	Ir 2·2	Pt 2·2	Au 2·4	Hg 1·9	Tl 1·8	Pb 1·8	Bi 1·9	Po 2·0	At 2·2	Rn
Fr 0·7	Ra 0·9	Ac 1·1															

Ce	Pr	Nd	Pm	Sm	Eu	Gd	Tb	Dy	Ho	Er	Tm	Yb	Lu
1·1	1·1	1·2	1·2	1·2	1·1	1·1	1·2	1·1	1·2	1·2	1·2	1·1	1·2

Th	Pa	U	Np	Pu	Am	Cm	Bk	Cf	Es	Fm	Md	No	Lr
1·3	1·5	1·7	1·3	1·3	1·3	1·3	1·3	1·3	1·3	1·3	1·3	1·3	

17 Percentage Ionic Character of a Single Bond

Pauling has estimated the amount of partial ionic character of a covalent bond in a binary compound empirically by means of the electronegativity difference of the two atoms which are bonded together: the approximate values obtained in this way are given in the following table.

Electronegativity difference	0·1	0·2	0·3	0·4	0·5	0·6	0·7	0·8	0·9	1·0	1·1	1·2	1·3	1·4	1·5	1·6
Percentage ionic character	0·5	1	2	4	6	9	12	15	19	22	26	30	34	39	43	47

Electronegativity difference	1·7	1·8	1·9	2·0	2·1	2·2	2·3	2·4	2·5	2·6	2·7	2·8	2·9	3·0	3·1	3·2
Percentage ionic character	51	55	59	63	67	70	74	76	79	82	84	86	88	89	91	92

18 Dipole Moments of Inorganic Compounds in the Vapour Phase

Many molecules have a permanent dipole moment μ: the debye, symbol D, is used as a unit for this physical quantity ($1 \text{ D} \cong 3.335\,640 \times 10^{-30}$ C m).

Compound	μ/D	Compound	μ/D
CH_4	0	N_2H_4	1·84
SiH_4	0	H_2O_2	2·13
NH_3	1·48	CO	0·10
PH_3	0·55	CO_2	0
AsH_3	0·16	N_2O	0·17
SbH_3	0·12	NO	0·16
H_2O	1·84	NO_2	0·4
H_2S	0·92	SO_2	1·63
H_2Se	0·4	CCl_4	0
H_2Te	0·2	NF_3	0·22
HF	1·91	PCl_3	0·78
HCl	1·05	CS_2	0
HBr	0·80	HCN	2·8
HI	0·42	$Ni(CO)_4$	0

19 Metallic, Covalent and van der Waals Radii

(a) denotes metallic radius divided by nanometre.
(b) denotes single bond covalent radius divided by nanometre.
The figure in parentheses is the van der Waals radius divided by nanometre.

	H	He
(a)	–	–
(b)	0·037	–
	(0·12)	

	Li	Be		B	C	N	O	F	Ne
(a)	0·152	0·112		–	–	–	–	–	–
(b)	0·123	0·089		0·080	0·077	0·074	0·074	0·072	–
					(0·15)	(0·140)	(0·135)	(0·160)

	Na	Mg		Al	Si	P	S	Cl	Ar
(a)	0·186	0·160		0·143	–	–	–	–	–
(b)	0·157	0·136		0·125	0·117	0·110	0·104	0·099	–
						(0·19)	(0·185)	(0·180)	(0·192)

	K	Ca	Sc	Ti	V	Cr	Mn	Fe	Co	Ni	Cu	Zn	Ga	Ge	As	Se	Br	Kr
(a)	0·231	0·197	0·160	0·146	0·131	0·125	0·129	0·126	0·125	0·124	0·128	0·133	0·141	–	–	–	–	–
(b)	0·203	0·174	0·144	0·132	0·122	0·117	0·117	0·116	0·116	0·115	0·117	0·125	0·125	0·122	0·121	0·117	0·114	–
															(0·20)	(0·200)	(0·195)	(0·197)

	Rb	Sr	Y	Zr	Nb	Mo	Tc	Ru	Rh	Pd	Ag	Cd	In	Sn	Sb	Te	I	Xe
(a)	0·244	0·215	0·180	0·157	0·141	0·136	0·135	0·133	0·134	0·138	0·144	0·149	0·166	0·162	–	–	–	–
(b)	0·216	0·191	0·162	0·145	0·134	0·129	–	0·124	0·125	0·128	0·134	0·141	0·150	0·140	0·141	0·137	0·133	–
															(0·22)	(0·220)	(0·215)	(0·217)

	Cs	Ba	La	Hf	Ta	W	Re	Os	Ir	Pt	Au	Hg	Tl	Pb	Bi	Po	At	Rn
(a)	0·262	0·217	0·188	0·157	0·143	0·137	0·137	0·134	0·135	0·138	0·144	0·152	0·171	0·175	0·170	0·14	–	–
(b)	0·235	0·198	0·169	0·144	0·134	0·130	0·128	0·126	0·126	0·129	0·134	0·144	0·155	0·154	0·152	–	0·140	–

	Fr	Ra	Ac
(a)	0·27	0·220	0·20
(b)	–	–	–

20 Ionic Radii

The data in the following table are ionic radii divided by nanometre: some of the values given are for purely hypothetical ions. The figures in parentheses indicate the charge on the ion.

										H 0·208(−1)	He
Li 0·060(+1)	Be 0·031(+2)										
Na 0·095(+1)	Mg 0·065(−2)										
K 0·133(+1)	Ca 0·099(+2)	Sc 0·081(+3)	Ti 0·090(+2) 0·068(+4)	Vª 0·074(+3) 0·059(+5)	Crᵇ 0·069(+3) 0·052(+6)	Mn 0·080(+2) 0·046(+7)	Fe 0·076(+2) 0·064(+3)	Co 0·078(+2) 0·063(+3)			
Rb 0·148(+1)	Sr 0·113(+2)	Y 0·093(+3)	Zr 0·080(−4)	Nb 0·070(+5)	Mo 0·068(−4) 0·062(−6)	Tc	Ru 0·069(+3) 0·065(+4)	Rh 0·086(+2)			
Cs 0·169(+1)	Ba 0·135(+2)	La 0·115(+3)	Hf 0·081(+4)	Ta 0·073(+5)	W 0·068(+4) 0·067(+6)	Re	Os 0·067(+4)	Ir 0·066(+4)			
Fr 0·176(−1)	Ra 0·140(−2)	Ac 0·118(−3)									

Ce 0·111(+3) 0·101(+4)	Pr 0·109(−3) 0·092(+4)	Nd 0·108(−3)	Pm 0·106(−3)	Sm 0·104(−3)	Eu 0·112(−2)	Gd 0·102(+3)
Th 0·114(−3) 0·095(+4)	Pa 0·112(−3) 0·091(+4)	U 0·111(−3) 0·089(+4)	Np 0·109(+3) 0·088(+4)	Pu 0·107(+3) 0·086(−4)	Am 0·106(+3) 0·085(−4)	Cm

NH_4^+
0·143

ª Also 0·088(+2)
ᵇ Also 0·084(+2)

			B 0·020(+3)	**C** 0·260(−4) 0·015(+4)	**N** 0·171(−3) 0·011(+5)	**O** 0·140(−2) 0·009(+6)	**F** 0·136(−1) 0·007(+7)	**Ne**
			Al 0·050(+3)	**Si** 0·271(−4) 0·041(+4)	**P** 0·212(−3) 0·034(+5)	**S** 0·184(−2) 0·029(+6)	**Cl** 0·181(−1) 0·026(+7)	**Ar**
Ni 0·078(+2) 0·062(+3)	**Cu** 0·096(+1) 0·069(+2)	**Zn** 0·074(+2)	**Ga** 0·148(+1) 0·062(+3)	**Ge** 0·093(+2) 0·053(+4)	**As** 0·222(−3) 0·047(+5)	**Se** 0·198(−2) 0·042(+6)	**Br** 0·195(−1) 0·039(+7)	**Kr**
Pd 0·050(+2)	**Ag** 0·126(+1)	**Cd** 0·097(+2)	**In** 0·132(+1) 0·081(+3)	**Sn** 0·112(+2) 0·071(+4)	**Sb** 0·245(−3) 0·062(+5)	**Te** 0·221(−2) 0·056(+6)	**I** 0·216(−1) 0·050(+7)	**Xe**
Pt 0·052(+2)	**Au** 0·137(+1)	**Hg** 0·110(+2)	**Tl** 0·140(+1) 0·095(+3)	**Pb** 0·120(+2) 0·084(+4)	**Bi** 0·120(+3) 0·074(+5)	**Po** 0·067(+6)	**At** 0·062(+7)	**Rn**

Tb 0·100(+3)	**Dy** 0·099(+3)	**Ho** 0·097(+3)	**Er** 0·096(+3)	**Tm** 0·095(+3)	**Yb** 0·113(+2)	**Lu** 0·093(+3)
Bk	**Cf**	**Es**	**Fm**	**Md**	**No**	**Lr**

21 The Shapes of Molecules and Ions

Shape		Examples
Linear	180°	$Cl-Be-Cl$ $O=C=O$ $H-C\equiv N$ $H-C\equiv C-H$ $[H_3N\rightarrow Ag.\leftarrow NH_3]^+$
Bent		$104.5°$ $H_2S\,(92°)$ $NO_2^-\,(115°)$ $ClO_2^-\,(110.5°)$
Trigonal planar	120°	NO_3^- CO_3^{2-} SO_3 C_2H_4
Pyramidal		$107°$ $PH_3(93°)$ ClO_3^- SO_3^{2-}
T-shape		
Tetrahedral	109°28'	NH_4^+ ClO_4^- SO_4^{2-} PO_4^{3-} $Ni(CO)_4$
Square planar		ICl_4^- $[Cu(H_2O)_4]^{2+}$
Trigonal bipyramidal	90° 120°	
Octahedral	90°	many six-coordinated metal complexes, e.g. $[Al(H_2O)_6]^{3+}$
Pentagonal bipyramidal	90° 72°	

22 Spatial Distribution of Hybrid Orbitals

Hybrid orbital	Distribution
sp	linear
sp^2	trigonal planar
sp^3	tetrahedral
sp^2d	square planar
sp^3d	trigonal bipyramidal
sp^3d^2	octahedral
sp^3d^3	pentagonal bipyramidal

23 Covalent Bond Lengths

l denotes covalent bond length.

Bond	l/nm	Bond	l/nm
H−H	0·074	C−H	0·109
C−C	0·154	Si−H	0·146
C=C	0·134	N−H	0·101
C≡C	0·120	P−H	0·142
C⋯C (in benzene)	0·139	O−H	0·096
Si−Si	0·235	S−H	0·135
N−N	0·146	F−H	0·092
N=N	0·120	Cl−H	0·128
N≡N	0·110	Br−H	0·141
P−P (P_4)	0·221	I−H	0·160
O−O	0·148		
O=O	0·121	C−O	0·143
S−S (S_8)	0·207	C=O	0·122
S=S	0·188	C⋯O (in phenol)	0·136
F−F	0·142	C−N	0·147
Cl−Cl	0·199	C=N	0·127
Br−Br	0·228	C≡N	0·116
I−I	0·267	C⋯N (in phenylamine)	0·135
		C−F	0·138
		C−Cl	0·177
		C⋯Cl (in chloro-benzene)	0·169
		C−Br	0·193
		C−I	0·214
		Si−O	0·150

24 Average Bond Enthalpies at 298 K

For a diatomic molecule, XY, the bond enthalpy ΔH is defined as the enthalpy change for the process:

$$X-Y(g) = X(g)+Y(g)$$

For a polyatomic molecule, XY_2, the *average* bond enthalpy is defined as half the enthalpy change for the process:

$$Y-X-Y(g) = X(g)+2Y(g)$$

Bond	ΔH/kJ mol^{-1}	Bond	ΔH/kJ mol^{-1}
H−H	436	C−H	412
D−D	442	Si−H	318
C−C	348	N−H	388
C=C	612	P−H	322
C≡C	837	O−H	463
C⋯C (benzene)	518	S−H	338
Si−Si	226	F−H	562
Ge−Ge	188	Cl−H	431
Sn−Sn	151	Br−H	366
N−N	163	I−H	299
N=N	409		
N≡N	944	C−O	360
P−P	172	C=O	743
O−O	146	C−N	305
O=O	496	C=N	613
S−S	264	C≡N	890
F−F	158	C−F	484
Cl−Cl	242	C−Cl	338
Br−Br	193	C−Br	276
I−I	151	C−I	238
		Si−O	374

25 Thermodynamic Data

$\Delta H_f^{\ominus}$ denotes the standard enthalpy of formation of a substance from its elements at 298 K and 101 325 N m^{-2} (1 atm) pressure.

$\Delta G_f^{\ominus}$ denotes the standard Gibbs free energy of formation of a substance from its elements at 298 K and 101 325 N m^{-2} (1 atm) pressure.

$S^{\ominus}$ denotes the standard entropy of a substance at 298 K and 101 325 N m^{-2} (1 atm) pressure.

$C_p^{\ominus}$ denotes the standard molar heat capacity of a substance at constant pressure at 298 K.

s denotes solid, l liquid, g gas, and am amorphous.

(a) Inorganic Compounds

Substance	State	$\Delta H_f^{\ominus}$ kJ mol^{-1}	$\Delta G_f^{\ominus}$ kJ mol^{-1}	$S^{\ominus}$ J K^{-1} mol^{-1}	$C_p^{\ominus}$ J K^{-1} mol^{-1}
Aluminium					
Al	s	0	0	28·3	24·3
Al$_2$O$_3$	s (α, corundum)	−1669	−1576	51·0	79·0
Al(OH)$_3$	am	−1273			
AlF$_3$	s	−1301	−1230	96	
AlCl$_3$	s	−695	−637	170	89·1
AlBr$_3$	s	−527	−505	180	102
AlI$_3$	s	−315	−314	200	99·2
Al$_4$C$_3$	s	−129	−121	100	
AlN	s	−241	−210	20	32
Al$_2$S$_3$	s	−509	−492	96	
Al(NO$_3$)$_3$.9H$_2$O	s	−3754	−2930	569	
Al$_2$(SO$_4$)$_3$	s	−3435	−3092	239	259
AlNH$_4$(SO$_4$)$_2$.12H$_2$O	s	−5939	−4933	697	683
AlK(SO$_4$)$_2$.12H$_2$O	s	−6057	−5137	687	
Antimony					
Sb	s	0	0	43·9	25·4
Sb$_2$O$_3$	s	−705	−623	123	101
Sb$_2$O$_4$	s	−895	−787	127	115
Sb$_2$O$_5$	s	−981	−839	125	118
SbF$_3$	s	−909			
SbCl$_3$	s	−382	−325	186	
SbCl$_5$	l	−438			
SbOCl	s	−380			
SbH$_3$	s	140			
Sb$_2$S$_3$	s (black)	−182	−180	127	
Sb$_2$S$_3$	am (orange)	−151			
Argon					
Ar	g	0	0	155	20·8
Arsenic					
As	s (α, grey)	0	0	35	25·0

Substance	State	$\Delta H_f^{\ominus}$ kJ mol^{-1}	$\Delta G_f^{\ominus}$ kJ mol^{-1}	$S^{\ominus}$ J K^{-1} mol^{-1}	$C_p^{\ominus}$ J K^{-1} mol^{-1}
As_4	g	149	105	290	
As_4O_6	s (octahedral)	-1314	-1152	214	191
As_2O_5	s	-915	-772	105	116
AsF_3	l	-949	-902	181	127
$AsCl_3$	l	-336	-295	234	
AsH_3	g	172			
As_2S_3	s	-150			
H_3AsO_4	s	-900			
Barium					
Ba	s	0	0	67	26·4
BaO	s	-558	-528	70·3	47·4
BaO_2	s	-630			
$Ba(OH)_2.8H_2O$	s	-3345			
BaF_2	s	-1201	-1148	96·2	71·2
$BaCl_2$	s	-860	-811	130	75·3
$BaCl_2.2H_2O$	s	-1461	-1296	203	155
$BaBr_2$	s	-755			
$BaBr_2.2H_2O$	s	-1365			
BaI_2	s	-602			
$BaI_2.2H_2O$	s	-1218			
BaH_2	s	-171			
Ba_3N_2	s	-364			
BaS	s	-444			
$BaCO_3$	s (witherite)	-1219	-1139	112	85·4
$Ba(NO_3)_2$	s	-992	-795	214	
$BaSO_4$	s	-1465	-1353	132	102
Beryllium					
Be	s	0	0	9·5	17·8
BeO	s	-611	-582	14·1	25·4
$Be(OH)_2$	s	-907			
$BeCl_2$	s	-512			
$BeBr_2$	s	-370			
BeI_2	s	-212			
Be_3N_2	s	-568	-512	33·4	
BeS	s	-234			
$BeSO_4$	s	-1197			
Bismuth					
Bi	s	0	0	56·9	26
Bi_2O_3	s	-577	-497	152	114
$Bi(OH)_3$	s	-710			
$BiCl_3$	s	-379	-319	190	
BiOCl	s	-365	-322	86·2	
Bi_2S_3	s	-183	-165	148	
Boron					
B	s	0	0	6·5	12·0
B_2O_3	s	-1264	-1184	54·0	62·3
BF_3	g	-1111	-1093	254	50·5
BCl_3	l	-418	-379	209	
BBr_3	l	-221	-219	229	

Substance	State	$\dfrac{\Delta H_f^{\ominus}}{\text{kJ mol}^{-1}}$	$\dfrac{\Delta G_f^{\ominus}}{\text{kJ mol}^{-1}}$	$\dfrac{S^{\ominus}}{\text{J K}^{-1}\text{mol}^{-1}}$	$\dfrac{C_P^{\ominus}}{\text{J K}^{-1}\text{mol}^{-1}}$
B_2H_6	g	31	82·8	233	56·4
BN	s	−134	−114	30	25
B_2S_3	s	−238			
H_3BO_3	s	−1089	−963	89·6	82·0
Bromine					
Br_2	l	0	0	152	71·6
Br_2	g	30·7	3·1	245	36·0
BrCl	g	14·7	−0·9	240	
HBr	g	−36·2	−53·2	198	29·1
Cadmium					
Cd	s (α)	0	0	51·5	25·9
CdO	s	−255	−225	54·8	43·4
$Cd(OH)_2$	s	−558	−471	95·4	
CdF_2	s	−690	−648	110	
$CdCl_2$	s	−389	−343	118	
$CdCl_2.2\cdot5H_2O$	s	−1129	−943	233	
$CdBr_2$	s	−314	−293	134	
$CdBr_2.4H_2O$	s	−1491	−1246	312	
CdI_2	s	−201	−201	168	
Cd_3N_2	s	162			
CdS	s	−144	−141	70	
$CdCO_3$	s	−748	−670	105	
$Cd(NO_3)_2.4H_2O$	s	−165			
$CdSO_4$	s	−926	−820	137	
Caesium					
Cs	s	0	0	82·8	31·0
Cs_2O	s	−318			
Cs_2O_2	s	−402			
CsOH	s	−407			
CsF	s	−531			
CsCl	s	−433			
CsBr	s	−394	−383	120	
CsI	s	−337	−334	130	
CsH	g	121	102	214	
Calcium					
Ca	s	0	0	41·6	26·3
CaO	s	−635	−604	40	42·8
$Ca(OH)_2$	s	−987	−897	76·2	84·5
CaF_2	s	−1214	−1162	68·9	67·0
$CaCl_2$	s	−795	−750	114	72·6
$CaCl_2.6H_2O$	s	−2608			
$CaBr_2$	s	−675	−656	130	
CaI_2	s	−535	−530	140	
CaH_2	s	−189	−150	40	
CaC_2	s	−62·8	−67·8	70·3	62·3
Ca_3N_2	s	−432	−369	100	
CaS	s	−483	−477	56·5	
$CaCO_3$	s (calcite)	−1207	−1129	92·9	81·9
$CaCO_3$	s (aragonite)	−1207	−1128	88·7	81·2

Substance	State	$\Delta H_f^{\ominus}$ kJ mol^{-1}	$\Delta G_f^{\ominus}$ kJ mol^{-1}	$S^{\ominus}$ J K^{-1} mol^{-1}	$C_p^{\ominus}$ J K^{-1} mol^{-1}
$Ca(HCO_3)_2$	s	-1354			
$Ca(NO_3)_2$	s	-937	-742	193	
$Ca(NO_3)_2.4H_2O$	s	-2132	-1701	340	
$Ca_3(PO_4)_2$	s (α)	-4126	-3890	241	232
$CaSO_4$	s (anhydrite)	-1433	-1320	107	99·6
$CaSO_4.0·5H_2O$	s (α)	-1575	-1435	130	120
	s (β)	-1573	-1434	134	124
$CaSO_4.2H_2O$	s	-2021	-1796	194	186
Carbon					
C	s (graphite)	0	0	5·7	8·6
C	s (diamond)	1·9	2·9	2·4	6·1
C	g	715	673	158	21
CO	g	-111	-137	198	29·1
CO_2	g	-394	-395	214	37·1
CF_4	g	-680	-661	262	
CCl_4	g	-107	$-64·0$	309	83·5
CCl_4	l	-139	$-68·6$	214	132
CBr_4	g	50·2	36	358	
$COCl_2$	g	-223	-210	289	60·7
CS_2	l	87·9	63·6	151	75·7
HCN	g	130	120	202	35·9
HCN	l	105	121	113	70·6
$(CN)_2$	g	308	296	242	56·9
Cerium					
Ce	s	0	0	57·7	25·9
Chlorine					
Cl_2	g	0	0	223	33·9
Cl_2O	g	76·2	93·7	266	
ClO_2	g	103	123	249	
Cl_2O_7	g	265			
ClF	g	$-55·6$	$-56·9$	218	
HCl	g	$-92·3$	$-95·3$	187	29·1
Chromium					
Cr	s	0	0	23·8	23·4
Cr_2O_3	s	-1128	-1047	81·2	119
CrO_3	s	-579			
$Cr(OH)_3$	s	-1033			
$CrCl_2$	s	-396	-356	115	70·6
$CrCl_3$	s	-563	-494	126	90·1
CrO_2Cl_2	l	-568			
Cobalt					
Co	s	0	0	28	25·6
CoO	s	-239	-213	43·9	
Co_3O_4	s	-879			
CoF_2	s	-665			
$CoCl_2$	s	-326	-282	106	78·7
$CoCl_2.6H_2O$	s	-2130			
$CoBr_2$	s	-232			

Substance	State	$\Delta H_f^{\ominus}$ kJ mol^{-1}	$\Delta G_f^{\ominus}$ kJ mol^{-1}	$S^{\ominus}$ J K^{-1} mol^{-1}	$C_p^{\ominus}$ J K^{-1} mol^{-1}
CoI_2	s	-102			
CoS	s	-84.5	-82.8	55	
$CoCO_3$	s	-722	-650		
$Co(NO_3)_2.6H_2O$	s	-2216			
$CoSO_4$	s	-868	-762	113	
$CoSO_4.7H_2O$	s	-2986			
Copper					
Cu	s	0	0	33.3	24.5
Cu_2O	s	-167	-146	101	69.9
CuO	s	-155	-127	43.5	44.4
$Cu(OH)_2$	s	-448			
CuF_2	s	-531			
$CuCl$	s	-135	-119	84.5	
$CuCl_2$	s	-206			
$CuCl_2.2H_2O$	s	-808			
$CuBr$	s	-105	-99.6	91.6	
$CuBr_2$	s	-141			
CuI	s	-67.8	-69.5	96.6	54.0
CuI_2	s	-7.1			
Cu_2S	s	-79.5	-86.2	121	76.3
CuS	s	-48.5	-49.0	66.5	47.8
$CuCO_3$	s	-595	-518	88	
$Cu(NO_3)_2$	s	-307			
$Cu(NO_3)_2.3H_2O$	s	-1219			
Cu_2SO_4	s	-750			
$CuSO_4$	s	-770	-662	113	
$CuSO_4.H_2O$	s	-1084	-917	150	
$CuSO_4.5H_2O$	s	-2278	-1880	305	281
Deuterium					
D_2	g	0	0	145	29.2
D_2O	g	-249	-235	198	34.3
D_2O	l	-295	-244	76.0	82.4
HDO	g	-246	-234	199	33.7
HDO	l	-290	-242	79.3	78.9
HD	g	0.2	-1.6	144	29.2
Fluorine					
F_2	g	0	0	203	31.5
F_2O	g	23	41	247	
HF	g	-269	-271	174	29.1
Gallium					
Ga	s	0	0	42.7	26.6
Ga_2O	s	-340			
Ga_2O_3	s	-1079			
$Ga(OH)_3$	s		-833		
$GaCl_3$	s	-525			
$GaBr_3$	s	-387			
GaI_3	s	-214			
GaN	s	-100			

Thermodynamic Data

Substance	State	$\Delta H_f^{\ominus}$ kJ mol^{-1}	$\Delta G_f^{\ominus}$ kJ mol^{-1}	$S^{\ominus}$ J K^{-1} mol^{-1}	$C_p^{\ominus}$ J K^{-1} mol^{-1}
Germanium					
Ge	s	0	0	42·4	26·1
GeO$_2$	am	−537			
GeCl$_4$	l	−544			
GeH$_4$	g			214	45·0
Gold					
Au	s	0	0	47·7	25·2
Au$_2$O$_3$	s	80·8	163	130	
AuCl	s	−35			
AuCl$_3$	s	−118			
Hafnium					
Hf	s	0	0	54·8	25·7
HfO$_2$	s	−1094			
Helium					
He	g	0	0	126	20·8
Hydrogen					
H$_2$	g	0	0	131	28·8
H	g	218	203	115	20·8
H$_2$O	g	−242	−229	189	33·6
H$_2$O	l	−286	−237	69·9	75·3
H$_2$O$_2$	g	−133			
H$_2$O$_2$	l	−188	−118	102	
Indium					
In	s	0	0	52·3	27·4
In$_2$O$_3$	s	−931			
In(OH)$_3$	s	−895	−762	100	
InCl	s	−186			
InCl$_3$	s	−537			
InBr$_3$	s	−404			
InI$_3$	s	−230			
InN	s	−20			
Iodine					
I$_2$	s	0	0	117	55·0
I$_2$	g	62·2	19·4	261	36·9
I$_2$O$_5$	s	−177			
ICl	g	17·6	−5·5	247	35·4
ICl$_3$	s	−88·3	−22·4	172	
IBr	g	40·8	3·8	259	
HI	g	25·9	1·3	206	29·2
HIO$_3$	s	−23·9			
Iridium					
Ir	s	0	0	36	25
IrO$_2$	s	−168			
IrCl$_2$	s	−179			
IrCl$_3$	s	−257			

Substance	State	$\Delta H_f^{\ominus}$ kJ mol^{-1}	$\Delta G_f^{\ominus}$ kJ mol^{-1}	$S^{\ominus}$ J K^{-1} mol^{-1}	$C_p^{\ominus}$ J K^{-1} mol^{-1}
Iron					
Fe	s	0	0	27·2	25·2
$Fe_{0.95}O$	s (wüstite)	−266	−244	54·0	
Fe_2O_3	s (haematite)	−822	−741	90·0	105
Fe_3O_4	s (magnetite)	−1117	−1015	146	
$Fe(OH)_2$	s	−568	−484	80	
$Fe(OH)_3$	s	−824			
$FeCl_2$	s	−341	−302	120	76·4
$FeCl_3$	s	−405			
$FeCl_3.6H_2O$	s	−2226			
$FeBr_2$	s	−251			
FeI_2	s	−125			
Fe_3C	s (cementite)	21	15	108	106
FeS	s (α)	−95·1	−97·6	67·4	54·8
FeS_2	s (pyrites)	−178	−167	53·1	61·9
$FeCO_3$	s (siderite)	−748	−674	92·9	82·1
$Fe(CO)_5$	l	−786			
$Fe(NO_3)_3.9H_2O$	s	−3282			
$FeSO_4$	s	−923	−815	108	
$FeSO_4.7H_2O$	s	−3007			
Krypton					
Kr	g	0	0	164	20·8
Lanthanum					
La	s	0	0	57·3	28
La_2O_3	s	−1916			
$LaCl_3$	s (α)	−1103			
Lead					
Pb	s	0	0	64·9	26·8
PbO	s (red)	−219	−189	67·8	
PbO	s (yellow)	−218	−188	69·4	49·0
PbO_2	s	−277	−219	76·6	64·4
Pb_3O_4	s	−735	−618	211	147
$Pb(OH)_2$	s	−515	−421	88	
PbF_2	s	−663	−620	120	
$PbCl_2$	s	−359	−314	136	77·0
$PbBr_2$	s	−277	−260	162	
PbI_2	s	−175	−174	177	
PbS	s	−94·3	−92·7	91·2	49·5
$PbCO_3$	s	−700	−626	131	
$Pb(CH_3COO)_2.3H_2O$	s	−1854			
$Pb(C_2H_5)_4$	l	220			
$Pb(NO_3)_2$	s	−449			
$PbSO_4$	s	−918	−811	147	104
Lithium					
Li	s	0	0	28·0	23·6
Li_2O	s	−596	−560	39·2	
Li_2O_2	s	−635			
LiOH	s	−487	−444	50	
LiF	s	−612	−584	35·9	

Substance	State	$\Delta H_f^{\ominus}$ kJ mol^{-1}	$\Delta G_f^{\ominus}$ kJ mol^{-1}	$S^{\ominus}$ J K^{-1} mol^{-1}	$C_p^{\ominus}$ J K^{-1} mol^{-1}
LiCl	s	-409			
LiBr	s	-350			
LiI	s	-271			
LiH	g	128	105	171	
Li$_3$N	s	-198			
Li$_2$CO$_3$	s	-1216	-1133	90·4	
LiNO$_3$	s	-482			
Li$_2$SO$_4$	s	-1434			
LiAlH$_4$	s	-101			
Magnesium					
Mg	s	0	0	32·5	23·9
MgO	s	-602	-570	27	37·4
Mg(OH)$_2$	s	-925	-834	63·1	
MgF$_2$	s	-1102	-1049	57·2	61·6
MgCl$_2$	s	-642	-592	89·5	
MgCl$_2$.6H$_2$O	s	-2500	-2116	366	316
MgBr$_2$	s	-518			
MgI$_2$	s	-360			
Mg$_3$N$_2$	s	-461			
MgS	s	-347			
MgCO$_3$	s	-1113	-1030	65·7	75·5
Mg(NO$_3$)$_2$	s	-790	-588	164	
Mg(NO$_3$)$_2$.6H$_2$O	s	-2612			
Mg$_3$(PO$_4$)$_2$	s	-4023			
MgNH$_4$PO$_4$.6H$_2$O	s	-3686			
MgSO$_4$	s	-1278	-1174	91·6	
MgSO$_4$.7H$_2$O	s	-3384			
Manganese					
Mn	s (α)	0	0	31·8	26·3
MnO	s	-385	-363	60·2	43·0
MnO$_2$	s	-521	-466	53·1	54·0
Mn$_2$O$_3$	s	-971			
Mn$_3$O$_4$	s	-1387	-1280	148	
Mn(OH)$_2$	am	-694	-610	88·3	
MnF$_2$	s	-791	-749	92·9	
MnCl$_2$	s	-482	-441	117	72·9
MnBr$_2$	s	-380			
MnI$_2$	s	-248			
MnS	s (green)	-204	-209	78·2	
MnS	s (red)	-199			
MnCO$_3$	s	-895	-818	85·8	
Mn(NO$_3$)$_2$	s	-696			
Mn(NO$_3$)$_2$.6H$_2$O	s	-2370			
MnSO$_4$	s	-1064	-956	112	
MnSO$_4$.4H$_2$O	s	-2256			
Mercury					
Hg	l	0	0	77·4	27·8
Hg	g	60·8	31·8	175	20·8
HgO	s (red)	$-90·7$	$-58·5$	72·0	45·7
HgO	s (yellow)	$-90·2$	$-58·4$	73·2	

Substance	State	$\Delta H_f^{\ominus}$ kJ mol^{-1}	$\Delta G_f^{\ominus}$ kJ mol^{-1}	$S^{\ominus}$ J K^{-1} mol^{-1}	$C_p^{\ominus}$ J K^{-1} mol^{-1}
Hg_2O	s	$-91 \cdot 2$	$-53 \cdot 6$		
Hg_2Cl_2	s	-265	-211	196	102
$HgCl_2$	s	-230	-177	121	$76 \cdot 6$
Hg_2Br_2	s	-207	-179	213	
$HgBr_2$	s	-170	-162	206	
Hg_2I_2	s (yellow)	-121	-113	239	106
HgI_2	s (red)	-105			
HgI_2	s (yellow)	-103			
HgS	s (red)	$-58 \cdot 2$	$-48 \cdot 8$	$77 \cdot 8$	
HgS	s (black)	$-54 \cdot 0$	$-46 \cdot 2$	$83 \cdot 3$	
$Hg_2(NO_3)_2.2H_2O$	s	-866			
$Hg(NO_3)_2.0 \cdot 5H_2O$	s	-389			
Hg_2SO_4	s	-742	-624	201	
$HgSO_4$	s	-704			
Molybdenum					
Mo	s	0	0	$28 \cdot 6$	$23 \cdot 5$
MoO_3	s	-754	-678	$78 \cdot 2$	$73 \cdot 6$
MoS_2	s	-232	-225	$63 \cdot 2$	
Neon					
Ne	g	0	0	146	$20 \cdot 8$
Nickel					
Ni	s	0	0	$30 \cdot 1$	$26 \cdot 0$
NiO	s	-244	-216	$38 \cdot 6$	$44 \cdot 4$
$Ni(OH)_2$	s	-538	-453	80	
NiF_2	s	-667			
$NiCl_2$	s	-316	-272	107	
$NiCl_2.6H_2O$	s	-2116	-1718	315	
$NiBr_2$	s	-227			
NiI_2	s	$-85 \cdot 8$			
NiS	s	$-73 \cdot 2$			
$NiCO_3$	s		-614		
$Ni(CO)_4$	g	-605	-587	402	
$Ni(NO_3)_2$	s	-428			
$Ni(NO_3)_2.6H_2O$	s	-2224			
$NiSO_4$	s	-891	-774	$77 \cdot 8$	
$NiSO_4.6H_2O$	s (blue)	-2688	-2222	306	340
Niobium					
Nb	s	0	0	35	25
Nb_2O_5	s	-1938			
Nitrogen					
N_2	g	0	0	192	$29 \cdot 1$
N_2O	g	$81 \cdot 6$	104	220	$38 \cdot 7$
NO	g	$90 \cdot 4$	$86 \cdot 7$	211	$29 \cdot 8$
N_2O_3	g	$92 \cdot 9$			
NO_2	g	$33 \cdot 9$	$51 \cdot 8$	240	$37 \cdot 9$
N_2O_4	g	$9 \cdot 7$	$98 \cdot 3$	304	$79 \cdot 1$
N_2O_5	g	15			
N_2O_5	s	$-41 \cdot 8$			

Substance	State	$\dfrac{\Delta H_f^{\ominus}}{\text{kJ mol}^{-1}}$	$\dfrac{\Delta G_f^{\ominus}}{\text{kJ mol}^{-1}}$	$\dfrac{S^{\ominus}}{\text{J K}^{-1}\text{mol}^{-1}}$	$\dfrac{C_p^{\ominus}}{\text{J K}^{-1}\text{mol}^{-1}}$
NF_3	g	−114			
NH_3	g	−46·2	−16·6	193	35·7
HN_3	g	294	328	237	
$NOCl$	g	52·6	66·4	264	
NH_4F	s	−467			
NH_4Cl	s	−315	−204	94·6	84·1
NH_4Br	s	−270			
NH_4I	s	−202			
NH_4HS	s	−159			
NH_4HCO_3	s	−852			
NH_4CNO	s	−312			
NH_4SCN	s	−83·7			
NH_4NO_2	s	−264			
NH_4NO_3	s	−365			182
$(NH_4)_3PO_4$	s	−1681			
$(NH_4)_2HPO_4$	s	−1573			
$NH_4H_2PO_4$	s	−1451	−1215	152	
$(NH_4)_2SO_4$	s	−1179	−900	220	187
NH_4HSO_4	s	−1024			143
$(NH_4)_2S_2O_8$	s	−1659			
NH_4VO_3	s	−1051	−886	141	
N_2H_4	l	50·4			
N_2H_5Cl	s	−196			
NH_2OH	s	−107			
NH_3OHCl	s	−310			
HNO_3	l	−173	−79·9	156	110
Osmium					
Os	s	0	0	33	25
OsO_4	s (white)	−384	−295	145	
OsO_4	s (yellow)	−391	−296	124	
Oxygen					
O_2	g	0	0	205	29·4
O_3	g	142	163	238	38·2
Palladium					
Pd	s	0	0	37	26
PdO	s	−85·4			
$PdCl_2$	s	−190			
Phosphorus					
P	s (white)	0	0	44·4	23·2
P	s (red)	−18			21
P	s (black)	−43·1			
P_4	g	54·9	24·4	280	
P_4O_6	s	−1640			
P_4O_{10}	s	−3012			
PCl_3	g	−306	−286	312	
PCl_3	l	−339			
PCl_5	g	−399	−325	353	
PCl_5	s	−463			
PBr_3	g	−150	−172	348	

Substance	State	$\Delta H_f^{\ominus}$ kJ mol^{-1}	$\Delta G_f^{\ominus}$ kJ mol^{-1}	$S^{\ominus}$ J K^{-1} mol^{-1}	$C_p^{\ominus}$ J K^{-1} mol^{-1}
PBr_3	l	-199			
PBr_5	s	-280			
PI_3	s	$-45 \cdot 6$			
$POCl_3$	g	-592	-545	325	
$POCl_3$	l	-632			
PH_3	g	$9 \cdot 2$	$18 \cdot 2$	210	
H_3PO_3	s	-972			
H_3PO_4	s	-1281			
HPO_3	s	-955			
Platinum					
Pt	s	0	0	$41 \cdot 8$	$26 \cdot 6$
$PtCl_2$	s	-148			
$PtCl_4$	s	-263			
Potassium					
K	s	0	0	$63 \cdot 6$	$29 \cdot 2$
K_2O	s	-362			
K_2O_2	s	-494			
K_2O_4	s	-561			
KOH	s	-426			
KF	s	-563	-533	$66 \cdot 6$	$49 \cdot 1$
KHF_2	s	-920	-852	104	$76 \cdot 9$
KCl	s	-436	-408	$82 \cdot 7$	$51 \cdot 5$
KBr	s	-392	-379	$96 \cdot 4$	$53 \cdot 6$
KI	s	-328	-322	104	$55 \cdot 1$
KH	g	126	105	198	
K_2S	s	-418			
K_2CO_3	s	-1146			
$KHCO_3$	s	-959			
KCN	s	-112			
KCNO	s	-412			
KSCN	s	-203			
KNO_2	s	-370			
KNO_3	s	-493	-393	133	$96 \cdot 3$
KNH_2	s	-118			
KH_2PO_4	s	-1569			
K_2SO_3	s	-1117			
K_2SO_4	s	-1434	-1316	176	130
$KHSO_4$	s	-1158			
$K_2S_2O_8$	s	-1917			
$KClO_3$	s	-391	-290	143	100
$KClO_4$	s	-434	-304	151	110
$KBrO_3$	s	-332	-244	149	105
KIO_3	s	-508	-426	152	106
K_2CrO_4	s	-1383			
$K_2Cr_2O_7$	s	-2033			
$KMnO_4$	s	-813	-714	172	119
$K_3Fe(CN)_6$	s	-173			
$K_4Fe(CN)_6$	s	-523			
Radium					
Ra	s	0	0	71	28
RaO	s	-523			

Substance	State	$\Delta H_f^{\ominus}$ kJ mol^{-1}	$\Delta G_f^{\ominus}$ kJ mol^{-1}	$S^{\ominus}$ J K^{-1} mol^{-1}	$C_p^{\ominus}$ J K^{-1} mol^{-1}
Radon					
Rn	g	0	0	176	20·8
Rhenium					
Re	s	0	0	42	26
Re_2O_7	s	−1245			
Rhodium					
Rh	s	0	0	32	26
RhO	s	−90·8			
Rh_2O_3	s	−286			
$RhCl_2$	s	−150			
$RhCl_3$	s	−230			
Rubidium					
Rb	s	0	0	69·4	30·4
Rb_2O	s	−330			
Rb_2O_2	s	−426			
RbOH	s	−414			
RbF	s	−549			
RbCl	s	−431	−412	119	51·5
RbBr	s	−389	−378	108	
RbI	s	−328	−326	118	
RbH	g	140			
Ruthenium					
Ru	s	0	0	29	23
RuO_2	s	−220			
$RuCl_3$	s	−260			
Scandium					
Sc	s	0	0	38	25
$ScCl_3$	s	−924			
Selenium					
Se	s (grey)	0	0	41·8	24·9
SeO_2	s	−230			
H_2Se	g	85·8	71·1	221	
SeF_6	g	−1030			
Silicon					
Si	s	0	0	18·7	19·9
SiO_2	s (quartz)	−859	−805	41·8	44·4
SiO_2	s (crystobalite)	−858	−804	42·6	44·2
SiO_2	s (tridymite)	−857	−803	43·3	44·4
SiF_4	g	−1550	−1510	284	76·2
$SiCl_4$	g	−610	−570	331	90·8
$SiCl_4$	l	−640	−573	239	145
$SiBr_4$	l	−398			
SiI_4	s	−132			
SiH_4	g	−61·9	−39	204	42·8
SiC	s	−112	−26·1	16·5	26·6

Substance	State	$\Delta H_f^{\ominus}$ kJ mol^{-1}	$\Delta G_f^{\ominus}$ kJ mol^{-1}	$S^{\ominus}$ J K^{-1} mol^{-1}	$C_p^{\ominus}$ J K^{-1} mol^{-1}
Silver					
Ag	s	0	0	42·7	25·5
Ag_2O	s	−30·6	−10·8	122	65·6
AgF	s	−203	−185	80	
AgCl	s	−127	−110	96·1	50·8
AgBr	s	−99·5	−93·7	107	52·4
AgI	s	−62·4	−66·3	114	54·4
Ag_2S	s (α)	−31·8	−40·2	146	
Ag_2CO_3	s	−506	−437	167	
AgCN	s	146	164	83·7	
AgSCN	s	87·9			
$AgNO_2$	s	−44·4	19·8	128	
$AgNO_3$	s	−123	−32·2	141	93·0
Ag_2SO_4	s	−713	−616	200	131
Ag_2CrO_4	s	−712	−622	217	
Sodium					
Na	s	0	0	51·0	28·4
Na_2O	s	−416	−377	72·8	68·2
Na_2O_2	s	−505			
NaOH	s	−427			80·3
NaF	s	−569	−541	58·6	46·0
$NaHF_2$	s	−906			
NaCl	s	−411	−384	72·4	49·7
NaBr	s	−360			52·3
NaI	s	−288			54·4
NaH	g	125	104	188	
NaH	s	−57·3			
Na_2S	s	−373			
$Na_2B_4O_7$	s	−3254			
$Na_2B_4O_7.10H_2O$	s	−6264			
Na_2CO_3	s	−1131	−1048	136	110
$Na_2CO_3.H_2O$	s	−1430			
$Na_2CO_3.10H_2O$	s	−4082			
$NaHCO_3$	s	−948	−852	102	87·6
NaCN	s	−89·8			
NaCNO	s	−400			
NaSCN	s	−175			
Na_2SiO_3	s	−1520	−1430	114	112
$NaNO_2$	s	−359			
$NaNO_3$	s	−467	−366	116	93·0
$NaNH_2$	s	−119			
Na_3PO_4	s	−1920			
$Na_3PO_4.12H_2O$	s	−5477			
Na_2HPO_4	s	−1747			
$Na_2HPO_4.12H_2O$	s	−5299			558
$NaNH_4HPO_4.4H_2O$	s	−2856			
Na_2SO_3	s	−1090	−1002	146	
$Na_2SO_3.7H_2O$	s	−3153			
Na_2SO_4	s	−1385	−1267	150	128
$Na_2SO_4.10H_2O$	s	−4324	−3644	593	587
$NaHSO_4$	s	−1126			
$Na_2S_2O_3$	s	−1117			

Substance	State	$\Delta H_f^\ominus$ kJ mol^{-1}	$\Delta G_f^\ominus$ kJ mol^{-1}	$S^\ominus$ J K^{-1} mol^{-1}	$C_p^\ominus$ J K^{-1} mol^{-1}
$Na_2S_2O_3.5H_2O$	s (I)	-2602			361
$NaClO_3$	s	-359			
$NaClO_4$	s	-386			101
Na_2CrO_4	s	-1329			
$NaBH_4$	s	-183	-120	105	86·6
Strontium					
Sr	s	0	0	54·4	25
SrO	s	-590	-560	54·4	
SrO_2	s	-643			
$Sr(OH)_2.8H_2O$	s	-3352			
SrF_2	s	-1214			
$SrCl_2$	s	-828	-781	120	
$SrCl_2.6H_2O$	s	-2624			
$SrBr_2$	s	-716			
SrI_2	s	-567			
SrH_2	s	-177			
Sr_3N_2	s	-391			
SrS	s	-452			
$SrCO_3$	s	-1219	-1138	97·1	
$Sr(NO_3)_2.4H_2O$	s	-2153			
$SrSO_4$	s	-1444	-1335	122	
Sulphur					
S	s (rhombic, α)	0	0	31·9	22·6
S	s (monoclinic, β)	0·3	0·1	32·6	23·6
S	g	223	182	168	23·7
S_8	g	102	49·8	430	
SO_2	g	-297	-300	248	39·8
SO_3	g	-395	-370	256	50·6
S_2Cl_2	l	$-60·2$			130
$SOCl_2$	l	-206			
SO_2Cl_2	l	-389			132
SF_6	g	-1100	-992	291	
H_2S	g	$-20·2$	$-33·0$	206	34·0
H_2SO_4	l	-811		138	
Tantalum					
Ta	s	0	0	41	25·3
Ta_2O_5	s	-2092	-1969	143	135
Technetium					
Tc	s	0	0	40	24
Tellurium					
Te	s	0	0	49·7	25·7
TeO_2	s	-325	-270	71·1	66·5
$TeCl_4$	s	-323			
TeF_6	g	-1320	-1220	338	
H_2Te	g	154	138	230	
Thallium					
Tl	s	0	0	64·4	26·6

Substance	State	$\Delta H_f^{\ominus}$ kJ mol^{-1}	$\Delta G_f^{\ominus}$ kJ mol^{-1}	$S^{\ominus}$ J K^{-1} mol^{-1}	$C_p^{\ominus}$ J K^{-1} mol^{-1}
Tl_2O	s	-175	-136	99·6	
TlOH	s	-238	-190	72·4	
$Tl(OH)_3$	s	-513			
TlCl	s	-205	-185	108	
$TlCl_3$	s	-251			
TlBr	s	-172	-166	111	
$TlBr_3$	s	-247			
TlI	s (II)	-124	-124	123	
Thorium					
Th	s	0	0	56·9	32
ThO_2	s	-1220			85·3
Tin					
Sn	s (white)	0	0	51·5	26·4
Sn	s (grey)	2·5	4·6	44·8	25·8
SnO	s	-286	-257	56·5	44·4
SnO_2	s	-581	-520	52·3	52·6
$Sn(OH)_2$	s	-579	-492	96·6	
$SnCl_2$	s	-350			
$SnCl_2.2H_2O$	s	-945			
$SnCl_4$	l	-545	-474	259	165
$SnBr_2$	s	-266			
SnI_2	s	-144			
SnS	s	$-77·8$	$-82·4$	98·7	
SnS_2	s	-167	-159	87·4	70·3
$Sn(SO_4)_2$	s	-1646			
Titanium					
Ti	s	0	0	30·3	25·2
TiO_2	s (rutile)	-912	-853	50·2	55·1
$TiCl_4$	l	-750	-674	253	157
Tungsten					
W	s	0	0	33	25·0
WO_3	s (yellow)	-840	-764	83·3	81·5
Uranium					
U	s	0	0	50·3	27·5
UO_2	s	-1130	-1080	77·8	
UO_3	s	-1260	-1180	98·6	
U_3O_8	s	-3760			
UF_4	s	-1850	-1760	151	118
UF_6	g	-2110	-2030	380	
Vanadium					
V	s	0	0	29·5	24·5
V_2O_3	s	-1210	-1130	98·7	
VO_2	s	-720	-665	51·6	59·4
V_2O_5	s	-1560	-1440	131	130
VCl_2	s	-452	-406	97·1	
VCl_3	s	-573	-502	131	

Substance	State	$\Delta H_f^{\ominus}$ kJ mol^{-1}	$\Delta G_f^{\ominus}$ kJ mol^{-1}	$S^{\ominus}$ J K^{-1} mol^{-1}	$C_p^{\ominus}$ J K^{-1} mol^{-1}
Xenon					
Xe	g	0	0	170	20·8
XeO$_3$	s	400			
XeF$_2$	g	$-82·0$			
XeF$_4$	s	-252	-123	145	118
XeF$_6$	g	-329			
Yttrium					
Y	s	0	0	46	26
YCl$_3$	s (γ)	-982			
Zinc					
Zn	s	0	0	41·6	25·1
ZnO	s	-348	-318	43·9	40·2
Zn(OH)$_2$	s	-642			
ZnCl$_2$	s	-416	-369	108	76·6
ZnBr$_2$	s	-327	-310	137	
ZnI$_2$	s	-209	-209	159	
ZnS	s (sphalerite)	-203	-198	57·7	45·2
ZnS	s (wurtzite)	-190			
ZnCO$_3$	s	-812	-731	82·4	
Zn(NO$_3$)$_2$	s	-482			
Zn(NO$_3$)$_2$.6H$_2$O	s	-2305			
ZnSO$_4$	s	-979	-872	125	
ZnSO$_4$.7H$_2$O	s	-3075	-2560	387	392
Zirconium					
Zr	s	0	0	38·4	26
ZrO$_2$	s	-1080	-1023	50·3	
ZrCl$_4$	s	-962	-874	186	

(b) Organic Compounds

Substance	State	$\Delta H_f^{\ominus}$ kJ mol^{-1}	$\Delta G_f^{\ominus}$ kJ mol^{-1}	$S^{\ominus}$ J K^{-1} mol^{-1}	$C_p^{\ominus}$ J K^{-1} mol^{-1}
CH$_4$	g	$-74·9$	$-50·8$	186	35·3
C$_2$H$_6$	g	$-84·7$	$-32·9$	230	52·7
C$_3$H$_8$	g	-104	$-23·5$	270	73·6
n-C$_4$H$_{10}$	g	-125	$-15·7$	310	97·5
n-C$_5$H$_{12}$	g	-146	$-8·2$	348	120
n-C$_6$H$_{14}$	g	-167	0·2	387	143
C$_2$H$_4$	g	52·3	68·1	219	43·5
C$_3$H$_6$	g	20·4	62·7	267	64·0
C$_4$H$_8$ (but-1-ene)	g	1·2	72·0	307	86
C$_4$H$_8$ (cis-but-2-ene)	g	$-5·7$	67·1	301	79
C$_4$H$_8$ (trans-but-2-ene)	g	$-10·1$	64·1	296	88
C$_2$H$_2$	g	227	209	201	43·9
C$_3$H$_4$	g	185	194	248	61
C$_4$H$_6$ (buta-1,3-diene)	g	112	152	279	79·5
C$_6$H$_{12}$ (cyclohexane)	l	-156	26·8	204	154

Substance	State	$\Delta H_f^{\ominus}$ kJ mol^{-1}	$\Delta G_f^{\ominus}$ kJ mol^{-1}	$S^{\ominus}$ J K^{-1} mol^{-1}	$C_p^{\ominus}$ J K^{-1} mol^{-1}
C_6H_6	g	82·9	130	269	81·6
C_6H_6	l	49·0	125	173	136
$C_6H_5CH_3$	g	50·0	122	320	104
$C_6H_5CH_2CH_3$	g	29·8	131	360	128
C_8H_8 (phenylethene)	g	148	214	345	
CH_3Cl	g	−82·0	−58·6	234	41
CH_2Cl_2	l	−117	−63·2	179	100
$CHCl_3$	l	−132	−71·6	203	116
CH_3Br	g	−36	−26	246	42·7
$CHBr_3$	l	−20	3	222	
CH_3I	l	−8·4	20	163	
CHI_3	s	141			
C_2H_5Cl	g	−105	−53·1	276	63
C_2H_5Br	l	−85·4			
C_2H_5I	l	−31			
$CH_2{=}CHCl$	g	31	52	264	
CH_2ClCH_2Cl	l	−166	−80·3	208	129
C_6H_5Cl	g	52·3	99	314	
$(CH_3)_2O$	g	−185	−114	267	66·1
CH_3OH	g	−201	−162	238	45·2
CH_3OH	l	−239	−166	127	81·6
C_2H_5OH	g	−235	−169	282	65
C_2H_5OH	l	−278	−175	161	111
C_6H_5OH	s	−163	−50·9	146	
$HCHO$	g	−116	−110	219	35
CH_3CHO	g	−166	−134	266	62·8
$(CH_3)_2CO$	g	−216	−152	295	74·9
HCO_2H	g	−363	−336	251	
HCO_2H	l	−409	−346	129	99·2
CH_3CO_2H	l	−487	−392	160	123
$C_6H_5CO_2H$	s	−385	−245	167	147
$CH_3CO_2C_2H_5$	l	−481			
CH_3COCl	l	−275	−208	201	
CH_3CONH_2	s	−320			
CH_3CN	l	53·1	100	144	
CH_3NH_2	g	−28	28	242	54·0
$C_2H_5NH_2$	g	−48·5	37	285	70
$CO(NH_2)_2$	s	−333	−47·1	105	93·3

26 Thermodynamic Data Relating to Change of State

$\Delta H_m^{\ominus}$ denotes the molar enthalpy of melting, i.e. for the transition solid → liquid.

$\Delta H_v^{\ominus}$ denotes the molar enthalpy of vaporization, i.e. for the transition liquid → vapour.

$\Delta H_s^{\ominus}$ denotes the molar enthalpy of sublimation, i.e. for the transition solid → vapour.

$\Delta H_a^{\ominus}$ denotes the enthalpy of atomization, i.e. the standard enthalpy of formation of the gaseous monatomic element at 298 K and 101 325 N m^{-2} (1 atm) pressure.

T_m denotes melting temperature.

T_b denotes boiling temperature.

Note. The ΔH values in part (a) of this table refer to 1 mol of *atoms* of the element.

(a) Elements

Element	T_m/K	$\Delta H_m^{\ominus}$/kJ mol^{-1}	T_b/K	$\Delta H_v^{\ominus}$/kJ mol^{-1}	$\Delta H_a^{\ominus}$/kJ mol^{-1}
H	14·0	0·059	20·5	0·45	218
He	3·5	0·02	4·3	0·084	0
Li	453	3·0	1600	136	161
Be	1550	12	2750	309	321
B	2600	22	4200	540	590
C (graphite)	4000		5100	715 (ΔH_s)	715
N	63·0	0·36	77·4	2·79	473
O	54·4	0·22	90·2	3·41	248
F	53·6	0·26	85·0	3·16	79·1
Ne	24·6	0·33	27·1	1·80	0
Na	371	2·6	1163	101	109
Mg	923	8·95	1380	132	150
Al	933	10·7	2740	284	314
Si	1680	46·4	2630	300	439
P (white)	317	0·63	553	12·4	315
S (β)	392	1·42	718	10	223
Cl	172	3·2	238	10·2	121
Ar	83·8	1·18	87·4	6·53	0
K	337	2·3	1047	79·1	90·0
Ca	1123	9·2	1760	153	193
Sc	1810	16	3000	310	340
Ti	1950	15	3530	427	469
V	2170	18	3300	444	515
Cr	2160	14	2755	347	398
Mn	1510	14·6	2370	225	279
Fe	1810	15·4	3200	354	418
Co	1765	15·2	3170	390	427
Ni	1726	17·6	3000	379	431
Cu	1356	13·0	2868	305	339
Zn	692	7·36	1180	115	130
Ga	303	5·61	2670	256	289
Ge	1210	32	3100	330	377
As	1090	27·7	886 (subl.)	32·4 (ΔH_s)	290

Element	T_m/K	$\Delta H_m^{\ominus}$/kJ mol^{-1}	T_b K	$\Delta H_v^{\ominus}$/kJ mol^{-1}	$\Delta H_a^{\ominus}$/kJ mol^{-1}
Se	490	5·23	958	14·0	202
Br	266	5·27	332	15·0	112
Kr	116	1·6	121	9·04	0
Rb	312	2·3	961	69·0	85·8
Sr	1041	9·2	1650	141	164
Y	1770	17	3200	390	431
Zr	2120	17	3850	502	611
Nb	2740	27	3600	694	724
Mo	2880	28	5800	536	651
Tc	2500	23	3800	502	648
Ru	2700	26	5200	619	640
Rh	2240	22	4800	531	556
Pd	1825	17	4250	380	390
Ag	1234	11·3	2480	254	289
Cd	594	6·11	1040	100	113
In	430	3·26	2300	225	244
Sn	505	7·20	2540	290	301
Sb	903	19·8	1650	195	254
Te	723	17·9	1260	49·8	199
I	387	7·87	457	22	107
Xe	161	2·30	165	12·6	0
Cs	302	2·09	963	66·1	78·7
Ba	987	7·66	1910	149	176
La	1190	11	3740	400	427
Hf	2500	22	5700	648	669
Ta	3250	28	5700	753	774
W	3680	33·7	6200	774	844
Re	3450	33	5900	636	791
Os	3300	27	5300	678	782
Ir	2710	28	5600	636	665
Pt	2042	22	4800	510	565
Au	1336	12·7	3240	342	369
Hg	234	2·3	630	58·2	60·8
Tl	577	4·27	1730	162	186
Pb	600	5·10	2017	177	196
Bi	544	11·0	1830	179	208
Po	527	13	1230	120	144
At	(575)	(12)	(650)	30	(92)
Rn	202	2·9	211	16·4	0
Fr	(300)	(2·1)	(950)	(64)	(73)
Ra	970	10	1410	115	130

(b) Compounds

Compound	T_m/K	$\Delta H_m^{\ominus}$/kJ mol^{-1}	T_b/K	$\Delta H_v^{\ominus}$/kJ mol^{-1}
CH_4	90·7	0·94	111·7	8·2
C_6H_6	279	9·83	353	30·8
CCl_4	250	2·5	350	30·5
CS_2	162	4·39	320	27·2
C_2H_5OH	156	4·60	352	43·5

Compound	T_m/K	$\Delta H_m^{\ominus}/\text{kJ mol}^{-1}$	T_b/K	$\Delta H_v^{\ominus}/\text{kJ mol}^{-1}$
NH_3	196	5·65	240	23·4
PCl_3	182	4·52	347	30·5
H_2O	273	6·02	373	41·1
H_2S	188	2·38	212	18·7
H_2Se	213	2·5	232	19·9
H_2Te	224	(4)	271	23·8
H_2O_2	273	12·2	423	43·1
SO_2	200	7·41	263	24·9
HF	190	3·93	293	7·5
HCl	158	1·99	188	16·2
HBr	185	2·41	206	17·6
HI	222	2·87	238	19·8
LiF	1140	26·8	1940	213
LiCl	880	13·4	1620	151
LiBr	820	12·1	1580	148
LiI	720	6·28	1450	171
NaF	1261	33·5	1970	201
NaCl	1074	28·9	1740	171
NaBr	1020	25·5	1660	159
NaI	920	22·2	1600	159
KF	1120	28·2	1800	161
KCl	1040	25·5	1680	162
KBr	1000	30	1710	156
KI	950	24	1600	145
RbCl	988	18·4	1650	154
CsCl	918	15·1	1560	149
AgCl	728	13·2	1840	178
$MgCl_2$	987	43·1	1680	137
$BaCl_2$	1230	22·6	1830	

27 Thermodynamics of Metal Extraction

The standard Gibbs free energy change $\Delta G^{\ominus}$ for the reaction of a series of elements with oxygen may be plotted against the temperature T, the equation for each reaction being written with one molecule of oxygen gas. This graphical representation of free energy data is known as an Ellingham diagram.

(Reference: H. J. T. Ellingham, *J. Soc. Chem. Ind.*, 1944, **63**, 125.)

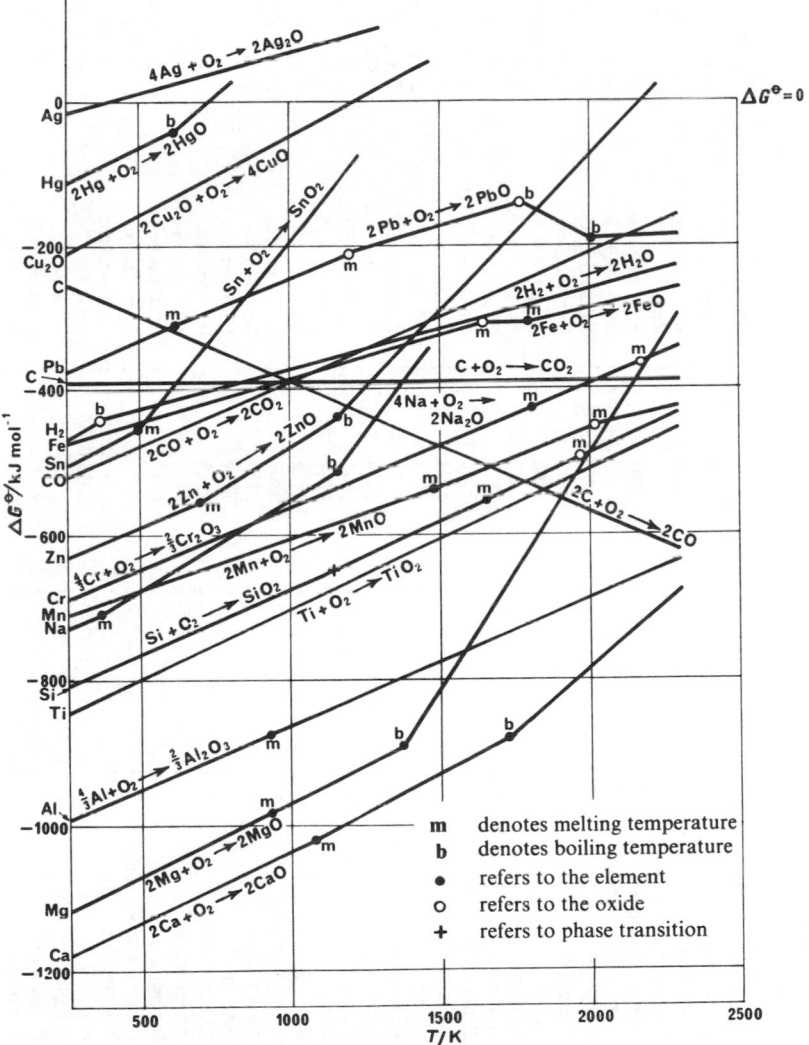

28 Enthalpies of Combustion

The values of the molar enthalpy of combustion ΔH in the following table refer to a temperature of 298 K and a pressure of 101 325 N m^{-2} (1 atm).

Substance	Formula	State	$-\Delta H/\text{kJ mol}^{-1}$	Substance	Formula	State	$-\Delta H/\text{kJ mol}^{-1}$
Hydrogen	H_2	g	285·8	Phenylmethanol	C_7H_7OH	l	4056
Sulphur (rhombic, α)	S	s	296·9	Cyclohexanol	$C_6H_{11}OH$	l	3727
Sulphur (monoclinic, β)	S	s	297·2	Phenol	C_6H_5OH	s	3064
Carbon (graphite)	C	s	393·5	Ethoxyethane	$(C_2H_5)_2O$	l	2727
Carbon (diamond)	C	s	395·4	Methanal	HCHO	g	561·1
Carbon monoxide	CO	g	283·0	Ethanal	CH_3CHO	l	1167
Methane	CH_4	g	890·4	Benzaldehyde	C_6H_5CHO	l	3520
Ethane	C_2H_6	g	1560	Propanone	$(CH_3)_2CO$	l	1786
Propane	C_3H_8	g	2220	Pentan-3-one	$(C_2H_5)_2CO$	l	3078
Butane	C_4H_{10}	g	2877	Phenylethanone	$CH_3COC_6H_5$	s	4138
Pentane	C_5H_{12}	l	3509	Diphenylmethanone	$(C_6H_5)_2CO$	s	6512
Hexane	C_6H_{14}	l	4194	Methanoic acid	HCO_2H	l	262·8
Octane	C_8H_{18}	l	5512	Ethanoic acid	CH_3CO_2H	l	876·1
Cyclohexane	C_6H_{12}	l	3924	Benzoic acid	$C_6H_5CO_2H$	s	3227
Ethene	C_2H_4	g	1409	Ethanedioic acid	$(CO_2H)_2$	s	246·4
Buta-1,3-diene	C_4H_6	g	2542	Benzoyl chloride	C_6H_5COCl	l	3275
Ethyne	C_2H_2	g	1299	Ethanoic anhydride	$(CH_3CO)_2O$	l	1807
Benzene	C_6H_6	l	3273	Ethyl ethanoate	$CH_3CO_2C_2H_5$	l	2246
Methylbenzene	C_7H_8	l	3909	Ethanamide	CH_3CONH_2	s	1182
Naphthalene	$C_{10}H_8$	s	5157	Benzamide	$C_6H_5CONH_2$	s	3546
Anthracene	$C_{14}H_{10}$	s	7114	Ethanonitrile	CH_3CN	l	1265
Chloroethane	C_2H_5Cl	g	1325	Benzonitrile	C_6H_5CN	l	3621
Bromoethane	C_2H_5Br	g	1425	Methylamine	CH_3NH_2	g	1072
Iodoethane	C_2H_5I	l	1490	Ethylamine	$C_2H_5NH_2$	g	1709
(Chloromethyl) benzene	C_7H_7Cl	l	3709	Phenylamine	$C_6H_5NH_2$	l	3397
Trichloromethane	$CHCl_3$	l	373·2	Nitrobenzene	$C_6H_5NO_2$	l	3094
Methanol	CH_3OH	l	715·0	Urea	$CO(NH_2)_2$	s	634·3
Ethanol	C_2H_5OH	l	1371	Glucose	$C_6H_{12}O_6$	s	2816
Propan-1-ol	C_3H_7OH	l	2010	Sucrose	$C_{12}H_{22}O_{11}$	s	5644
Butan-1-ol	C_4H_9OH	l	2673				

29 Enthalpies of Hydrogenation of Gases

The values of the molar enthalpy of hydrogenation ΔH in the following table refer to gases at 355 K and 101 325 N m^{-2} (1 atm) pressure.
(Reference: G. B. Kistiakowsky *et al.*, *J. Chem. Soc.*, 1935, **57**, 65 et seq.)

Compound	$-\Delta H$/kJ. mol^{-1}	Compound	$-\Delta H$/kJ mol^{-1}
Ethene, $CH_2 = CH_2$	157·3	Cyclohexene, $CH_2CH_2CH = CHCH_2CH_2$	119·6
Propene, $CH_3CH = CH_2$	126·0		
2-Methylpropene, $(CH_3)_2C = CH_2$	118·8	1,3-Cyclohexadiene,	
But-1-ene, $C_2H_5CH = CH_2$	126·9	$CH_2CH = CHCH = CHCH_2$	231·7
But-2-ene, $CH_3CH = CHCH_3$ (cis-)	119·5	Benzene, C_6H_6	208·4
		Ethylbenzene, $C_6H_5CH_2CH_3$	204·7
But-2-ene, $CH_3CH = CHCH_3$ (trans-)	115·6	1,2-Dimethylbenzene, $C_6H_4(CH_3)_2$	197·7
		1,3,5-Trimethylbenzene. $C_6H_3(CH_3)_3$	199·2
Buta-1,3-diene, $CH_2 = CH - CH = CH_2$	238·8	Phenylethene, $C_6H_5CH = CH_2$	324·2
Ethyne, $CH \equiv CH$	314·0	Cyclopentadiene, $CH = CHCH_2CH = CH$	212·8
Propyne, $CH_3C \equiv CH$	291·6	Ethanal, $CH_3CH = O$	70·1
But-2-yne, $CH_3C \equiv CCH_3$	274·4	Propanone, $(CH_3)_2C = O$	56·1

30 Enthalpies of Solution at 298 K

The values of the molar enthalpy of solution ΔH in the following table refer to the formation of an 'infinitely dilute' aqueous solution.

				$\Delta H/\text{kJ mol}^{-1}$				
Anion	OH^-	F^-	Cl^-	Br^-	I^-	CO_3^{2-}	NO_3^-	SO_4^{2-}
Cation								
Li^+	−21·2	+4·5	−37·2	−49·1	−63·3	−17·6	−2·7	−30·2
Na^+	−42·7	·0·3	+3·9	−0·6	−7·6	−24·6	+20·5	−2·3
K^+	−55·2	−17·7	+17·2	+20·0	+20·5	32·6	+34·9	+23·8
Rb^+	−62·6	−26·3	+16·7	+21·9	+26·1	−40·3	+36·7	+24·3
Cs^+	−71·0	−45·9	+17·9	+25·9	+33·2	·52·8	+39·9	+17·1
Ag^+		−20·3	+65·7	+84·5	+112	+41·8	+22·5	+17·6
NH_4^+		+5·0	+15·2	+16·2	+13·4		+25·8	+6·2
Mg^{2+}	+2·8	−17·7	−155	−186	−214	−25·3	−85·5	−91·2
Ca^{2+}	−16·2	+13·4	−82·9	−110	−120	·12·3	−18·9	−17·8
Sr^{2+}	−46·0	+10·9	−52·0	−71·6	−90·4	−3·4	+17·7	−8·7
Ba^{2+}	−51·8	+3·8	−13·2	−25·4	−47·7	+4·2	+40·4	+19·4
Cu^{2+}	+48·1	−63·2	−51·5	−35·2	−26·4	−16·7	−41·8	−73·3
Zn^{2+}	+29·9		−71·5	−67·2	−55·2	−16·2	−83·9	−81·4
Cd^{2+}		−40·7	−18·4	−2·8	+14·0		−33·7	−53·7
Pb^{2+}	+56·1	+6·3	+25·9	+36·8	+64·8	+22·2	+37·7	+11·3
Al^{3+}		−209	−332	−360	−378			−318

31 Enthalpies of Neutralization at 298 K

The values of the molar enthalpy of neutralization ΔH in the following table refer to infinitely dilute aqueous solution.

Acid	*Base*	$-\Delta H/\text{kJ mol}^{-1}$
HCl	NaOH	57·1
HCl	KOH	57·2
HNO_3	NaOH	57·3
HNO_3	KOH	57·3
HNO_3	$\frac{1}{2}Ba(OH)_2$	58·2
HF	NaOH	68·6
HCl	NH_3	52·2
CH_3CO_2H	NaOH	55·2
HCN	KOH	11·7
HCN	NH_3	5·4

32 Enthalpies and Gibbs Free Energies of Formation of Aqueous Ions

The following table gives the standard enthalpy of formation $\Delta H_f^{\ominus}$ and the standard Gibbs free energy of formation $\Delta G_f^{\ominus}$ of various aqueous ions at 'infinite dilution', *from their elements*, relative to that of $H^+(aq)$ taken as zero.

Ion	$\Delta H_f^{\ominus}/kJ$ mol^{-1}	$\Delta G_f^{\ominus}/kJ$ mol^{-1}	Ion	$\Delta H_f^{\ominus}/kJ$ mol^{-1}	$\Delta G_f^{\ominus}/kJ$ mol^{-1}
H^+	0	0	OH^-	−229·9	−157·3
Li^+	−278·5	−293·8	F^-	−329·1	−276·5
Na^+	−239·6	−261·9	Cl^-	−167·5	−131·2
K^+	−251·2	−282·3	Br^-	−120·9	−102·9
Rb^+	−248·5	−282·2	I^-	−55·9	−51·7
Cs^+	−247·6	−282·0	I_3^-	−51·9	−51·5
NH_4^+	−132·8	−79·5	S^{2-}	+32·6	+86·2
Be^{2+}	−389·0	−329·2	HS^-	−17·7	+12·6
Mg^{2+}	−461·9	−456·0	CO_3^{2-}	−676·2	−528·1
Ca^{2+}	−542·9	−553·0	HCO_3^-	−691·1	−587·1
Sr^{2+}	−545·5	−557·3	CN^-	+151·0	+165·7
Ba^{2+}	−538·3	−560·7	HCO_2^-	−409·9	−334·6
Al^{3+}	−524·7	−481·2	$CH_3CO_2^-$	−488·8	−369·4
Sn^{2+}	−10·0	−26·2	$C_2O_4^{2-}$	−824·2	−674·9
Pb^{2+}	+1·6	−24·3	NO_2^-	−106·3	−34·5
Cr^{2+}	−139	−165	NO_3^-	−206·6	−110·6
Cr^{3+}	−256	−205	PO_4^{3-}	−1284	−1025
Mn^{2+}	−218·7	−223·3	HPO_4^{2-}	−1299	1094
Fe^{2+}	−87·9	−84·9	$H_2PO_4^-$	−1302	−1135
Fe^{3+}	−47·7	−10·5	SO_3^{2-}	−624·2	−497·1
Co^{2+}	−67·4	−51·5	HSO_3^-	−628·0	−527·3
Ni^{2+}	−64·0	−46·4	SO_4^{2-}	−907·5	−742·0
Cu^+	+51·9	+50·2	HSO_4^-	−885·7	−752·9
Cu^{2+}	+64·4	+65·0	$S_2O_3^{2-}$	−644	−532
Ag^+	+105·9	+77·1	$S_4O_6^{2-}$	−1210	−1016
Zn^{2+}	−152·4	−147·2	ClO_3^-	−98·3	−2·5
Cd^{2+}	−72·4	−77·7	BrO_3^-	−83·6	+1·7
Hg^{2+}	+174	+164·8	IO_3^-	−221	−128
$[Fe(CN)_6]^{4-}$	+496·6	+772·4	ClO_4^-	−131·4	−10·3
$[Fe(CN)_6]^{3-}$	+561·5	+806·3	CrO_4^{2-}	−863·2	−706·2
$[Cu(NH_3)_4]^{2+}$	−339	−132	$Cr_2O_7^{2-}$	−1460	−1257
$[Ag(NH_3)_2]^+$	−111·8	−17·4	MnO_4^-	−518·4	−425·0

33 Hydration Enthalpies of Ions

The hydration enthalpy ΔH_{hyd} of an ion refers to the process:

$$M^{z\pm}(g) + aq = M^{z\pm}(aq)$$

Cation	$-\Delta H_{hyd}/kJ\ mol^{-1}$	Anion	$-\Delta H_{hyd}/kJ\ mol^{-1}$
H^+	(1091)	OH^-	460
Li^+	519	F^-	506
Na^+	406	Cl^-	364
K^+	322	Br^-	335
Rb^+	301	I^-	293
Cs^+	276		
Ag^+	464		
NH_4^+	301		
Mg^{2+}	1920		
Ca^{2+}	1650		
Sr^{2+}	1480		
Ba^{2+}	1360		
Fe^{2+}	1950		
Cu^{2+}	2100		
Zn^{2+}	2050		
Al^{3+}	4690		
Fe^{3+}	4430		

34 Lattice Enthalpies at 298 K (experimental and theoretical values)

Lattice enthalpy ΔH relates to the endothermic process

$$MX(s) = M^+(g) + X^-(g)$$

in which the gaseous ions of a crystal are separated to an infinite distance from each other.

(a) Experimental values

The data in the following tables are experimental values obtained by means of a suitable Born-Haber cycle.

(i) Alkali Metal Halides

	ΔH/kJ mol^{-1}			
	F	Cl	Br	I
Li	1022	846	800	744
Na	902	771	733	684
K	801	701	670	629
Rb	767	675	647	609
Cs	716	645	619	585

(ii) Other Substances

	ΔH/kJ mol^{-1}		ΔH/kJ mol^{-1}
CaF_2	2602	MgS	3238
$BeCl_2$	3006	CaS	2966
$MgCl_2$	2493	SrS	2779
$CaCl_2$	2237	BaS	2643
$SrCl_2$	2112	CuCl	976
$BaCl_2$	2018	AgF	955
MgO	3889	AgCl	905
CaO	3513	AgBr	890
SrO	3310	AgI	876
BaO	3152	NH_4Cl	640
		SnO_2	11770

(b) Theoretical values

Lattice energies may be calculated from electrostatic principles on the basis of a purely ionic model for the crystal. The Born-Mayer expression, which was used to calculate the values in the following tables, includes a constant A (known as the Madelung constant) the value of which depends on the geometry of the crystal. The expression derived by Kapustinskii does not require any knowledge of the structure of the compound concerned.

(Reference: A. F. Kapustinskii, *Quart. Rev.*, 1956, **10**(3), 283 *et seq.*)

(*i*) *Alkali Metal Halides*

	$\Delta H/kJ\ mol^{-1}$			
	F	Cl	Br	I
Li	1004	833	787	728
Na	891	766	732	686
K	795	690	665	632
Rb	761	674	644	607
Cs	728	636	611	582

(*ii*) *Other Substances*

	$\Delta H/kJ\ mol^{-1}$		$\Delta H/kJ\ mol^{-1}$
CaF_2	2611	AgF	870
MgO	3929	AgCl	770
CaO	3477	AgBr	758
SrO	3205	AgI	736
BaO	3042		

35 Solubility of Gases in Water

(a) Absorption Coefficients

The absorption coefficient of a gas is defined as the volume of gas, measured at $0\,°C$ and $101\,325\ N\ m^{-2}$ (1 atm) pressure, which will dissolve in unit volume of solvent at a stated temperature under a partial pressure of $101\,325\ N\ m^{-2}$ (1 atm) of the gas.

Gas	*Absorption coefficient at 0 °C*
Ammonia	1300
Argon	0·056
Carbon dioxide	1·71
Carbon monoxide	0·035
Chlorine	4·61
Ethene	0·25
Helium	0·0094
Hydrogen	0·021
Hydrogen chloride	506
Hydrogen sulphide	4·68
Nitrogen	0·024
Oxygen	0·049
Sulphur dioxide	79·8

(b) At Various Temperatures

The solubilities in the following table are given as the mass of gas per unit mass of water at a *total* pressure of 101 325 N m^{-2} (1 atm) (sum of the partial pressures of gas and water).

θ_C denotes Celsius temperature.

Gas	$\theta_C/°C$ 0	20	40	60	70	80	90
Ammonia	0·895	0·531	0·307				
Carbon dioxide	$3·35 \times 10^{-3}$	$1·69 \times 10^{-3}$	$9·73 \times 10^{-4}$	$5·76 \times 10^{-4}$			
Chlorine		$7·0 \times 10^{-3}$	$4·1 \times 10^{-3}$	$2·5 \times 10^{-3}$			
Hydrogen	$1·92 \times 10^{-6}$	$1·60 \times 10^{-6}$	$1·38 \times 10^{-6}$	$1·18 \times 10^{-6}$	$1·02 \times 10^{-6}$	$7·9 \times 10^{-7}$	$4·6 \times 10^{-7}$
Hydrogen chloride	0·823	0·721	0·633	0·561			
Hydrogen sulphide	$7·07 \times 10^{-3}$	$3·85 \times 10^{-3}$	$2·36 \times 10^{-3}$	$1·48 \times 10^{-3}$	$1·10 \times 10^{-3}$	$7·65 \times 10^{-4}$	$4·1 \times 10^{-4}$
Nitrogen	$2·94 \times 10^{-5}$	$1·90 \times 10^{-5}$	$1·39 \times 10^{-5}$	$1·05 \times 10^{-5}$	$8·5 \times 10^{-6}$	$6·6 \times 10^{-6}$	$3·8 \times 10^{-6}$
Oxygen	$6·94 \times 10^{-5}$	$4·34 \times 10^{-5}$	$3·08 \times 10^{-5}$	$2·27 \times 10^{-5}$	$1·86 \times 10^{-5}$	$1·38 \times 10^{-5}$	$7·9 \times 10^{-6}$
Sulphur dioxide	0·228	0·113	0·0541				

36 Solubility of Inorganic Compounds

The solubilities in the following table are given as 100 times the mass of anhydrous solute per unit mass of water, at 20 °C, unless otherwise stated.

vs denotes very soluble   s denotes soluble   ss denotes sparingly soluble   dec. denotes decomposes

	Anion	O^{2-}	OH^-	S^{2-}	F^-	Cl^-	Br^-	I^-	CO_3^{2-}	NO_3^-	SO_4^{2-}
	Cation										
Group I	Li^+	dec.	12·8	vs	$0·27^a$	83	177	165	1·33	70	35
	Na^+	dec.	109	19	4·0	36	91	179	21 (HCO_3^-, 9·6)	87	19·4
	K^+	dec.	112	s	95	34·7	67	144	112 (HCO_3^-, 22·4)	31·6	11·1
	Rb^+	dec.	177	vs	131	91	110	152	450	53	48
	Cs^+	dec.	330	vs	370	186	108	79	vs	23	179
	NH_4^+			vs	vs	37	75	172	12 (HCO_3^-)	192	75
Group II	Be^{2+}	ss	ss	dec.	vs	vs	s	dec.		107	39
	Mg^{2+}	ss	0·0009	dec.	0·008	54·2	102	148	ss	70	33
	Ca^{2+}	dec.	0·156	dec.	0·0016	74·5	142	209	ss	129	0·21
	Sr^{2+}	dec.	0·80	dec.	0·012	53·8	100	178	ss	71	0·013
	Ba^{2+}	dec.	3·9	dec.	0·12	36	104	205	ss	8·7	0·00024
Group III	Al^{3+}	ss	ss	dec.	0·55	70 dec.	dec.	dec.			38
	Ga^{3+}	ss	ss	dec.	0·002	vs	s	dec.		63	vs
	In^+	ss				dec.	dec.				
	In^{3+}	ss	ss	ss	0·04	vs	vs	dec.			s
	Tl^+	dec.	$25·9^b$	0·02	$78·6^c$	0·33	0·05	0·0006	$4·0^c$	9·55	4·87
	Tl^{3+}	ss		ss	dec.	vs	s	s		s	

^{a} At 18 °C ^{b} At 0 °C ^{c} At 15 °C

	Anion / Cation	O^{2-}	OH^-	S^{2-}	F^-	Cl^-	Br^-	I^-	CO_3^{2-}	NO_3^-	SO_4^{2-}
Group IV	Ge^{2+}	ss		0·24	s	dec.	dec.	s			
	Ge^{4+}	0·41		0·45dec.	dec.	dec.	dec.	dec.			33[b]
	Sn^{2+}	ss	ss	ss	s	270[a] dec.	s	0·98		dec.	
	Sn^{4+}	ss		ss	vs	dec.	dec.	dec.			ss
	Pb^{2+}	0·0017	0·016	ss	0·064	0·99	0·844	0·063	0·00011	55	
	Pb^{4+}	ss				dec.					
Group V	As^{3+}	3·7		ss	dec.	dec.	dec.	6·0[b]			
	As^{5+}	150[c]		ss							
	Sb^{3+}	ss		ss	dec.	dec.	dec.	dec.		dec.	ss
	Sb^{5+}	ss		ss		dec.					
	Bi^{3+}	ss	0·00014	ss	ss	dec.	dec.	ss		dec.	dec.
Transition elements	Cr^{3+}	ss	ss	dec.	ss	ss	ss			s	dec.
	Mn^{2+}	ss	ss	ss	1·05	72·3[b]	vs	s	0·0065	vs	63
	Fe^{2+}	ss	ss	ss	ss	64·4[d]	115	s	0·006	s	s
	Fe^{3+}	ss	ss	dec.	ss	dec.	s	s		s	dec.
	Co^{2+}	ss	ss	ss	1·5[b]	64	s	vs	ss	vs	36·2
	Ni^{2+}	ss	ss	ss	4[b]	64·2	120	130	0·009	vs	37
	Cu^+	ss		ss	ss	0·006	ss	0·0008[c]			dec.
	Cu^{2+}	ss	ss	ss	4·7	73	vs		ss	122	20·5
	Zn^{2+}	ss	ss	ss	1·62	432[b]	447	432[e]	0·001[a]	117	54
	Ag^+	0·0013		ss	195[e]	ss	ss	ss	0·0032	217	0·8
	Cd^{2+}	ss	ss	ss	4·3	140	98	84	ss	150	76
	Hg_2^{2+}	ss		ss	dec.	0·0002	4×10^{-6}	ss	ss	dec.	0·06
	Hg^{2+}	0·005		ss	dec.	6·9	0·55	0·01	ss	vs	dec.

[a] At 15 °C   [b] At 25 °C   [c] At 16 °C   [d] At 10 °C   [e] At 18 °C

37 Azeotropes: Binary Mixtures at Atmospheric Pressure

Azeotropic mixtures are also known as constant boiling temperature mixtures.

$\theta_{C,b}$ denotes the boiling temperature at 101 325 N m^{-2} pressure.

Components	$\theta_{c,b}$/°C	Azeotrope Per cent composition by mass	$\theta_{c,b}$/°C
Benzene	80·1	67·6	67·8
Ethanol	78·5	32·4	
Benzene	80·1	91·1	69·4
Water	100·0	8·9	
Ethanoic acid	118·1	3·0	76·6
Water	100·0	97·0	
Ethanoic acid	118·1	2·0	80·1
Benzene	80·1	98·0	
Ethanol	78·5	95·6	78·2
Water	100·0	4·4	
Hydrogen bromide	− 67·0	47·5	126·0
Water	100·0	52·5	
Hydrogen chloride	−83·7	20·2	108·6
Water	100·0	79·8	
Hydrogen fluoride	19·4	35·6	111·4
Water	100·0	64·4	
Hydrogen iodide	− 35·5	57·0	127·0
Water	100·0	43·0	
Nitric acid	86·0	68·0	120·5
Water	100·0	32·0	
Propan-1-ol	97·2	71·8	88·1
Water	100·0	28·2	
Propanone	56·2	20·0	64·7
Trichloromethane	61·2	80·0	
Propanone	56·2	88·5	56·1
Water	100·0	11·5	
Tetrachloromethane	76·8	84·2	65·0
Ethanol	78·5	15·8	
Tetrachloromethane	76·8	95·9	66·8
Water	100·0	4·1	
Trichloromethane	61·2	93·0	59·4
Ethanol	78·5	7·0	
Trichloromethane	61·2	97·0	56·3
Water	100·0	3·0	

38 Vapour Pressure of Water at Various Temperatures

p denotes vapour pressure and θ_C denotes Celsius temperature.

(a) 10 to 24·5 °C

$\theta_C/°C$	$p/kN\ m^{-2}$	$\theta_C/°C$	$p/kN\ m^{-2}$
10	1·23	17·5	2·00
10·5	1·27	18	2·06
11	1·31	18·5	2·13
11·5	1·36	19	2·19
12	1·40	19·5	2·27
12·5	1·45	20	2·34
13	1·50	20·5	2·41
13·5	1·55	21	2·49
14	1·60	21·5	2·56
14·5	1·65	22	2·64
15	1·70	22·5	2·72
15·5	1·76	23	2·81
16	1·82	23·5	2·90
16·5	1·88	24	2·98
17	1·94	24·5	3·07

(b) 0 to 100 °C

$\theta_C/°C$	$p/kN\ m^{-2}$	$\theta_C/°C$	$p/kN\ m^{-2}$
0	0·610	75	38·53
5	0·872	80	47·35
10	1·23	85	57·81
15	1·70	90	70·10
20	2·34	91	72·80
25	3·17	92	75·60
30	4·24	93	78·47
35	5·62	94	81·45
40	7·38	95	84·51
45	9·58	96	87·68
50	12·35	97	90·94
55	15·74	98	94·30
60	19·92	99	97·76
65	25·05	100	101·32
70	31·16		

39 Eutectic Mixtures

The *eutectic temperature* $\theta_{C,E}$ is the lowest temperature at which both the solid components of a mixture are in equilibrium with the liquid phase. $\theta_{C,m}$ denotes melting temperature.

Component 1	$\theta_{C,m}/°C$	Component 2	$\theta_{C,m}/°C$	$\theta_{C,E}/°C$	Composition of eutectic mixture (per cent by mass)	
Sn	232	Pb	327	183	Sn, 63·0	Pb, 37·0
Sn	232	Zn	420	198	Sn, 91·0	Zn, 9·0
Sn	232	Ag	961	221	Sn, 96·5	Ag, 3·5
Sn	232	Cu	1083	227	Sn, 99·2	Cu, 0·8
Sn	232	Bi	271	140	Sn, 42·0	Bi, 58·0
Sb	630	Pb	327	246	Sb, 12·0	Pb, 88·0
Bi	271	Pb	327	124	Bi, 55·5	Pb, 44·5
Bi	271	Cd	321	146	Bi, 60·0	Cd, 40·0
Cd	321	Zn	420	270	Cd, 83·0	Zn, 17·0

40 Transition Temperatures

$\theta_{C,t}$ denotes transition temperature.

Substance	System	$\theta_{C,t}/°C$
Sulphur	Rhombic (α) $\rightleftharpoons$ Monoclinic (β)	95·6
Tin	Grey (α) $\rightleftharpoons$ White (β)	13·2
Iron	α (body-centred cubic) $\rightleftharpoons$ γ (face-centred cubic)	906
	γ (face-centred cubic) $\rightleftharpoons$ δ (body-centred cubic)	1401
Sodium sulphate	$Na_2SO_4.10H_2O \rightleftharpoons Na_2SO_4 + 10H_2O$	32·4
Mercury(II) iodide	Tetragonal (red) $\rightleftharpoons$ Orthorhombic (yellow)	126
Ammonium chloride	α (CsCl structure) $\rightleftharpoons$ β (NaCl structure)	184
Caesium chloride	CsCl structure $\rightleftharpoons$ NaCl structure	445
Copper(I) mercury(II) iodide	Tetragonal (red) $\rightleftharpoons$ Cubic (dark brown)	69

41 Typical Equilibrium Data

The equilibrium constant K of the reaction:

$$aA + bB \ldots \ldots \rightleftharpoons lL + mM + \ldots \ldots$$

may be expressed in terms of concentrations, when

$$K_c = \frac{[L]_e^l [M]_e^m \cdots}{[A]_e^a [B]_e^b \cdots}$$

or in terms of partial pressures, when

$$K_p = \frac{p_L^l \, p_M^m \cdots}{p_A^a \, p_B^b \cdots}$$

(a) The Equilibrium $H_2(g) + I_2(g) \rightleftharpoons 2HI(g)$ at 763·8 K

(Reference: A. H. Taylor and R. H. Crist, *J. Amer. Chem. Soc.*, 1941, **63**, 1377.)

$[H_2]_e$/mol m^{-3}	$[I_2]_e$/mol m^{-3}	$[HI]_e$/mol m^{-3}	$K_c = \dfrac{[HI]_e^2}{[H_2]_e[I_2]_e}$	
2·265	2·840	17·15	45·7	
1·920	3·634	17·80	45·4	
1·699	4·057	17·79	45·7	
2·179	2·870	16·90	45·7	
2·484	2·514	16·95	46·0	(i)
2·636	2·305	16·64	45·7	
4·173	1·185	14·94	45·1	
3·716	1·478	15·76	45·2	
2·594	2·597	17·63	46·3	
1·894	1·896	12·83	45·9	
1·971	1·981	13·42	46·1	(ii)
2·413	2·424	16·41	46·0	

(i) Equilibrium established by combination of hydrogen and iodine.
(ii) Equilibrium established by decomposition of hydrogen iodide.

(b) The Equilibrium $C_2H_5OH + CH_3CO_2H \rightleftharpoons CH_3CO_2C_2H_5 + H_2O$ at 373 K

n denotes amount of substance.

(Reference: M. Berthelot and P. de Saint-Gilles, *Ann. Chim. Phys.*, 1862, **65**, 385.)

$\dfrac{n \text{ (alcohol)}}{n \text{ (acid)}}$	$\dfrac{n \text{ (ester)}}{\text{mol}}$	$K_c = \dfrac{[Ester][Water]}{[Alcohol][Acid]}$
0·18	0·17!	3·92
0·33	0·293	3·28
0·50	0·420	3·80
1·00	0·667	4·00
2·00	0·858	4·54
8·00	0·966	3·90

(c) The Synthesis of Ammonia $N_2 + 3H_2 \rightleftharpoons 2NH_3$ with a 3:1 mole ratio of H_2 to N_2

p denotes pressure and θ_C denotes Celsius temperature.

	Amount per cent of ammonia at equilibrium				
			$\theta_C/°C$		
$p/\text{MN m}^{-2}$	*300*	*400*	*500*	*600*	*700*
0·101 (1 atm)	2·18	0·44	0·13	0·05	0·02
1·01 (10 atm)	14·7	3·85	1·21	0·49	0·23
3·04 (30 atm)	31·8	10·7	3·62	1·43	0·66
10·1 (100 atm)	51·2	25·1	10·4	4·52	2·18
20·3 (200 atm)	62·8	36·3	17·6	8·20	4·10
101 (1000 atm)	92·6	79·8	57·5	31·4	12·9

42 Equilibrium Constants for Gaseous Reactions at Various Temperatures

The variation of equilibrium constant with temperature may be represented by the following equation due to van't Hoff:

$$\ln \frac{K_1}{K_2} = \frac{-\Delta H^\ominus}{R} \left(\frac{1}{T_1} - \frac{1}{T_2} \right)$$

$$\text{or, } \lg \frac{K_1}{K_2} = \frac{-\Delta H^\ominus}{2 \cdot 303 R} \left(\frac{1}{T_1} - \frac{1}{T_2} \right)$$

where K_1 and K_2 denote the equilibrium constants at temperatures T_1 and T_2 respectively, $\Delta H^\ominus$ denotes the standard enthalpy change of reaction and R denotes the gas constant. In addition to values of the standard enthalpy change at 298 K, values of the standard Gibbs free energy change $\Delta G^\ominus_{298}$ and the standard entropy change $\Delta S^\ominus_{298}$ are given for each reaction.

(a) The Equilibrium $N_2O_4(g) \rightleftharpoons 2NO_2(g)$

$$\Delta H^{\ominus}_{298} = +58{\cdot}1 \text{ kJ}$$
$$\Delta G^{\ominus}_{298} = +5{\cdot}3 \text{ kJ}$$
$$\Delta S^{\ominus}_{298} = +177 \text{ J K}^{-1}$$

T/K	K_p/atm
298	0·115
350	3·89
400	47·9
450	$3{\cdot}47 \times 10^2$
500	$1{\cdot}70 \times 10^3$
550	$6{\cdot}03 \times 10^3$
600	$1{\cdot}78 \times 10^4$

(b) The Equilibrium $N_2(g) + 3H_2(g) \rightleftharpoons 2NH_3(g)$

$$\Delta H^{\ominus}_{298} = -92{\cdot}4 \text{ kJ}$$
$$\Delta G^{\ominus}_{298} = -33{\cdot}2 \text{ kJ}$$
$$\Delta S^{\ominus}_{298} = -199 \text{ J K}^{-1}$$

T/K	K_p/atm^{-2}
298	$6{\cdot}76 \times 10^5$
400	40·7
500	$3{\cdot}55 \times 10^{-2}$
600	$1{\cdot}66 \times 10^{-3}$
700	$7{\cdot}76 \times 10^{-5}$

(c) The Equilibrium $2SO_2(g) + O_2(g) \rightleftharpoons 2SO_3(g)$

$$\Delta H^{\ominus}_{298} = -196 \text{ kJ}$$
$$\Delta G^{\ominus}_{298} - -140 \text{ kJ}$$
$$\Delta S^{\ominus}_{298} = -188 \text{ J K}^{-1}$$

T/K	K_p/atm^{-1}
298	$4{\cdot}0 \times 10^{24}$
500	$2{\cdot}5 \times 10^{10}$
700	$3{\cdot}0 \times 10^4$

(d) The Equilibrium $H_2(g) + I_2(g) \rightleftharpoons 2HI(g)$

$$\Delta H^{\ominus}_{298} = -10{\cdot}4 \text{ kJ}$$
$$\Delta G^{\ominus}_{298} = -16{\cdot}8 \text{ kJ}$$
$$\Delta S^{\ominus}_{298} = +21{\cdot}5 \text{ J K}^{-1}$$

T/K	K_p
298	$7{\cdot}94 \times 10^2$
500	$1{\cdot}60 \times 10^2$
700	54
764	46*

* Experimental value (see Table 41)

43 Standard Electrode Potentials

Standard electrode potentials $E^\ominus$ refer to half-reactions at 298 K in which all ions taking part in the electrode process are at unit activity, all gases are at $101\,325\,\mathrm{N\,m^{-2}}$ (1 atm) pressure and solids are in their most stable form at 298 K. As a reference standard, the half-reaction represented by the equation

$$H^+(a=1)+e^- = \tfrac{1}{2}H_2(101\,325\ \mathrm{N\ m^{-2}}\ \text{or}\ 1\ \text{atm})$$

is arbitrarily assigned a potential of zero at 298 K.

The sign convention used in the following table is that in which the half-reaction is written as a reduction process, i.e.

$$\text{oxidant}+ne^- = \text{reductant}$$

so that reactions which proceed to the right more readily than the H^+, H_2 reaction are given a positive potential and those which proceed less readily are given a negative potential; the positive electrode of a cell then has the more positive (or less negative) electrode potential. The electrode reactions are given in alphabetical order according to the element being reduced.

Electrode reaction	$E^\ominus/V$
$Ag^+ + e^- = Ag$	$+0\cdot80$
$Ag^{2+} + e^- = Ag^+$	$+1\cdot98$
$AgBr + e^- = Ag + Br^-$	$+0\cdot07$
$AgCN + e^- = Ag + CN^-$	$-0\cdot04$
$Ag(CN)_2^- + e^- = Ag + 2CN^-$	$-0\cdot38$
$AgCl + e^- = Ag + Cl^-$	$+0\cdot22$
$AgI + e^- = Ag + I^-$	$-0\cdot15$
$Ag(NH_3)_2^+ + e^- = Ag + 2NH_3$	$+0\cdot37$
$\tfrac{1}{2}Ag_2O + \tfrac{1}{2}H_2O + e^- = Ag + OH^-$	$+0\cdot34$
$Al^{3+} + 3e^- = Al$	$-1\cdot66$
$Al(OH)_4^- + 3e^- = Al + 4OH^-$	$-2\cdot35$
$As + 3H^+ + 3e^- = AsH_3$	$-0\cdot38$
$H_3AsO_4 + 2H^+ + 2e^- = H_3AsO_3 + H_2O$	$+0\cdot56$
$Au^+ + e^- = Au$	$+1\cdot68$
$Au^{3+} + 3e^- = Au$	$+1\cdot50$
$H_3BO_3 + 3H^+ + 3e^- = B + 3H_2O$	$-0\cdot73$
$Ba^{2+} + 2e^- = Ba$	$-2\cdot90$
$Be^{2+} + 2e^- = Be$	$-1\cdot85$
$BiO^+ + 2H^+ + 3e^- = Bi + H_2O$	$+0\cdot28$
$\tfrac{1}{2}Br_2 + e^- = Br^-$	$+1\cdot07$
$HOBr + H^+ + e^- = \tfrac{1}{2}Br_2 + H_2O$	$+1\cdot59$
$BrO_3^- + 6H^+ + 5e^- = \tfrac{1}{2}Br_2 + 3H_2O$	$+1\cdot52$
$CO_2 + H^+ + e^- = \tfrac{1}{2}H_2C_2O_4$	$-0\cdot49$
$Ca^{2+} + 2e^- = Ca$	$-2\cdot87$
$Cd^{2+} + 2e^- = Cd$	$-0\cdot40$
$Ce^{4+} + e^- = Ce^{3+}$ (in 1 mol dm^{-3} H$_2$SO$_4$)	$+1\cdot45$

Electrode reaction	$E^{\ominus}/V$
$\frac{1}{2}Cl_2 + e^- = Cl^-$	$+1\cdot36$
$HOCl + H^+ + e^- = \frac{1}{2}Cl_2 + H_2O$	$+1\cdot64$
$ClO_3^- + 6H^+ + 5e^- = \frac{1}{2}Cl_2 + 3H_2O$	$+1\cdot47$
$Co^{2+} + 2e^- = Co$	$-0\cdot28$
$Co^{3+} + e^- = Co^{2+}$	$+1\cdot82$
$Co(NH_3)_6^{2+} + 2e^- = Co + 6NH_3$	$-0\cdot43$
$Cr^{2+} + 2e^- = Cr$	$-0\cdot91$
$Cr^{3+} + 3e^- = Cr$	$-0\cdot74$
$Cr^{3+} + e^- = Cr^{2+}$	$-0\cdot41$
$\frac{1}{2}Cr_2O_7^{2-} + 7H^+ + 3e^- = Cr^{3+} + \frac{7}{2}H_2O$	$+1\cdot33$
$Cs^+ + e^- = Cs$	$-2\cdot92$
$Cu^+ + e^- = Cu$	$+0\cdot52$
$Cu^{2+} + 2e^- = Cu$	$+0\cdot34$
$Cu^{2+} + e^- = Cu^+$	$+0\cdot15$
$Cu^{2+} + I^- + e^- = CuI$	$+0\cdot86$
$Cu(NH_3)_4^{2+} + 2e^- = Cu + 4NH_3$	$-0\cdot05$
$D^+ + e^- = \frac{1}{2}D_2$	$-0\cdot003$
$\frac{1}{2}F_2 + e^- = F^-$	$+2\cdot87$
$Fe^{2+} + 2e^- = Fe$	$-0\cdot44$
$Fe^{3+} + 3e^- = Fe$	$-0\cdot04$
$Fe^{3+} + e^- = Fe^{2+}$	$+0\cdot77$
$Fe(CN)_6^{3-} + e^- = Fe(CN)_6^{4-}$	$+0\cdot36$
$Fe(OH)_3 + e^- = Fe(OH)_2 + OH^-$	$-0\cdot56$
$FeO_4^{2-} + 8H^+ + 3e^- = Fe^{3+} + 4H_2O$	$+2\cdot20$
$H^+ + e^- = \frac{1}{2}H_2$	$0\cdot00$
$\frac{1}{2}H_2 + e^- = H^-$	$-2\cdot25$
$\frac{1}{2}Hg_2^{2+} + e^- = Hg$	$+0\cdot79$
$Hg^{2+} + 2e^- = Hg$	$+0\cdot85$
$Hg^{2+} + e^- = \frac{1}{2}Hg_2^{2+}$	$+0\cdot91$
$Hg_2Cl_2 + 2e^- = 2Hg + 2Cl^-$	$+0\cdot27$
$\frac{1}{2}I_2 + e^- = I^-$	$+0\cdot54$
$HOI + H^+ + e^- = \frac{1}{2}I_2 + H_2O$	$+1\cdot45$
$IO_3^- + 6H^+ + 5e^- = \frac{1}{2}I_2 + 3H_2O$	$+1\cdot19$
$K^+ + e^- = K$	$-2\cdot92$
$Li^+ + e^- = Li$	$-3\cdot04$
$Mg^{2+} + 2e^- = Mg$	$-2\cdot38$
$Mn^{2+} + 2e^- = Mn$	$-1\cdot18$
$Mn^{3+} + e^- = Mn^{2+}$	$+1\cdot51$
$MnO_2 + 4H^+ + 2e^- = Mn^{2+} + 2H_2O$	$+1\cdot23$
$MnO_4^- + e^- = MnO_4^{2-}$	$+0\cdot56$
$MnO_4^- + 4H^+ + 3e^- = MnO_2 + 2H_2O$	$+1\cdot67$
$MnO_4^- + 8H^+ + 5e^- = Mn^{2+} + 4H_2O$	$+1\cdot52$
$\frac{1}{2}N_2 + 4H^+ + 3e^- = NH_4^+$	$+0\cdot27$
$HNO_2 + H^+ + e^- = NO + H_2O$	$+0\cdot99$
$NO_3^- + 2H^+ + e^- = NO_2 + H_2O$	$+0\cdot81$
$NO_3^- + 3H^+ + 2e^- = HNO_2 + H_2O$	$+0\cdot94$
$NO_3^- + 4H^+ + 3e^- = NO + 2H_2O$	$+0\cdot96$
$NO_3^- + 5H^+ + 4e^- = \frac{1}{2}N_2O + \frac{5}{2}H_2O$	$+1\cdot11$
$NO_3^- + 6H^+ + 5e^- = \frac{1}{2}N_2 + 3H_2O$	$+1\cdot24$
$NO_3^- + 10H^+ + 8e^- = NH_4^+ + 3H_2O$	$+0\cdot87$
$Na^+ + e^- = Na$	$-2\cdot71$
$Ni^{2+} + 2e^- = Ni$	$-0\cdot25$
$Ni(NH_3)_6^{2+} + 2e^- = Ni + 6NH_3$	$-0\cdot51$
$\frac{1}{2}H_2O_2 + H^+ + e^- = H_2O$	$+1\cdot77$

Electrode reaction	$E^{\ominus}/V$
$\frac{1}{2}O_2 + 2H^+ + 2e^- = H_2O$	$+1 \cdot 23$
$O_2 + 2H^+ + 2e^- = H_2O_2$	$+0 \cdot 68$
$\frac{1}{2}O_2 + H_2O + 2e^- = 2OH^-$	$+0 \cdot 40$
$O_3 + 2H^+ + 2e^- = O_2 + H_2O$	$+2 \cdot 07$
$P + 3H^+ + 3e^- = PH_3$	$-0 \cdot 04$
$H_3PO_4 + 2H^+ + 2e^- = H_3PO_3 + H_2O$	$-0 \cdot 28$
$Pb^{2+} + 2e^- = Pb$	$-0 \cdot 13$
$Pb^{4+} + 2e^- = Pb^{2+}$	$+1 \cdot 69$
$PbO_2 + 4H^+ + 2e^- = Pb^{2+} + 2H_2O$	$+1 \cdot 47$
$PbO_2 + H_2O + 2e^- = PbO + 2OH^-$	$+0 \cdot 28$
$Ra^{2+} + 2e^- = Ra$	$-2 \cdot 92$
$Rb^+ + e^- = Rb$	$-2 \cdot 92$
$S + 2e^- = S^{2-}$	$-0 \cdot 51$
$S + 2H^+ + 2e^- = H_2S$	$+0 \cdot 14$
$SO_4^{2-} + 4H^+ + 2e^- = H_2SO_3 + H_2O$	$+0 \cdot 17$
$\frac{1}{2}S_2O_8^{2-} + e^- = SO_4^{2-}$	$+2 \cdot 01$
$\frac{1}{2}S_4O_6^{2-} + e^- = S_2O_3^{2-}$	$+0 \cdot 09$
$Sb + 3H^+ + 3e^- = SbH_3$	$-0 \cdot 51$
$SbO^+ + 2H^+ + 3e^- = Sb + H_2O$	$+0 \cdot 21$
$Sn^{2+} + 2e^- = Sn$	$-0 \cdot 14$
$Sn^{4+} + 2e^- = Sn^{2+}$	$+0 \cdot 15$
$Sr^{2+} + 2e^- = Sr$	$-2 \cdot 89$
$Ti^{2+} + 2e^- = Ti$	$-1 \cdot 63$
$Ti^{3+} + e^- = Ti^{2+}$	$-0 \cdot 37$
$V^{2+} + 2e^- = V$	$-1 \cdot 2$
$V^{3+} + e^- = V^{2+}$	$-0 \cdot 26$
$VO^{2+} + 2H^+ + e^- = V^{3+} + H_2O$	$+0 \cdot 34$
$VO_2^+ + 2H^+ + e^- = VO^{2+} + H_2O$	$+1 \cdot 00$
$XeO_3 + 6H^+ + 6e^- = Xe + 3H_2O$	$+1 \cdot 8$
$Zn^{2+} + 2e^- = Zn$	$-0 \cdot 76$
$Zn(NH_3)_4^{2+} + 2e^- = Zn + 4NH_3$	$-1 \cdot 03$
$Zn(OH)_4^{2-} + 2e^- = Zn + 4OH^-$	$-1 \cdot 22$

44 Standard Reference Electrodes

The variation in the potential of a calomel electrode (saturated KCl) with temperature θ_C is given by the equation

$$E_{cal} = [0 \cdot 2444 - 0 \cdot 0007 \, (\theta_C/°C - 25)] \text{ V}$$

$\theta_C/°C$	$E^{\ominus}_{cal}/V$ (0·1 mol dm^{-3} KCl)	$E^{\ominus}_{cal}/V$ (1 mol dm^{-3} KCl)	$E^{\ominus}_{cal}/V$ (saturated KCl)
18	0·336	0·280	0·249
20	0·335	0·280	0·248
25	0·335	0·278	0·244
30	0·335	0·277	0·240
35	0·334	0·277	0·236
40	0·333	0·277	0·232

45 Dissociation Constants of Inorganic Acids

The dissociation constant of an acid K_a may conveniently be expressed in terms of the pK_a value where $pK_a = -\log_{10}(K_a/\text{mol dm}^{-3})$. The values given in the following table are for aqueous solutions at 298 K: the pK_1, pK_2 and pK_3 values refer to the first, second and third ionizations respectively.

Name	Formula	pK_a
Aluminium ion (hydrated)	$[Al(H_2O)_6]^{3+}$	4·9 (pK_1)
Ammonium ion	NH_4^+	9·25
Arsenic(III) acid	H_3AsO_3	9·22 (pK_1)
Arsenic(V) acid	H_3AsO_4	2·30 (pK_1)
Boric acid	H_3BO_3	9·24 (pK_1)
Bromic(I) acid	HOBr	8·70
Carbonic acid	H_2CO_3	$\begin{cases} 6·38^a \ (pK_1) \\ 10·32 \ (pK_2) \end{cases}$
Chloric(I) acid	HOCl	7·43
Chloric(III) acid	$HClO_2$	2·0
Chromium(III) ion (hydrated)	$[Cr(H_2O)_6]^{3+}$	3·9 (pK_1)
Hydrazinium ion	$N_2H_5^+$	7·93
Hydrocyanic acid	HCN	9·40
Hydrofluoric acid	HF	3·25
Hydrogen peroxide	H_2O_2	11·62 (pK_1)
Hydrogen sulphide	H_2S	$\begin{cases} 7·05 \ (pK_1) \\ 12·92 \ (pK_2) \end{cases}$
Hydroxyammonium ion	NH_3OH^+	5·82
Iodic(I) acid	HOI	10·52
Iodic(V) acid	HIO_3	0·8
Iron(III) ion (hydrated)	$[Fe(H_2O)_6]^{3+}$	2·22 (pK_1)
Lead(II) ion (hydrated)	$[Pb(H_2O)_n]^{2+}$	7·8 (pK_1)
Nitrous acid	HNO_2	3·34
Phosphinic acid	H_3PO_2	2·0
Phosphoric(V) acid	H_3PO_4	$\begin{cases} 2·15 \ (pK_1) \\ 7·21 \ (pK_2) \\ 12·36 \ (pK_3) \end{cases}$
Phosphonic acid	H_3PO_3	$\begin{cases} 2·00 \ (pK_1) \\ 6·58 \ (pK_2) \end{cases}$
Silicic acid	H_2SiO_3	$\begin{cases} 9·9 \ (pK_1) \\ 11·9 \ (pK_2) \end{cases}$
Sulphuric acid	H_2SO_4	1·92 (pK_2)
Sulphurous acid	H_2SO_3	$\begin{cases} 1·92 \ (pK_1) \\ 7·21 \ (pK_2) \end{cases}$

^{a} Some of the unionized acid exists as dissolved CO_2 molecules rather than H_2CO_3: pK_1 for the molecular species H_2CO_3 is approximately 3·7.

46 Ionic Product of Water at Various Temperatures

Ionic product $K_w = [H^+][OH^-]$.
θ_C denotes Celsius temperature.

	$\theta_C/°C$											
	0	5	10	15	20	25	30	35	40	45	50	100
Ionic product $K_w \times 10^{14}/\text{mol}^2\,\text{dm}^{-6}$	0·114	0·186	0 293	0·452	0·681	1·008	1·471	2·088	2·916	4·016	5·476	51·3

47 Solubility Products

For a sparingly soluble electrolyte M_xA_y in contact with its saturated solution, the solubility product is given by

$$K_{sp} = [M^{y+}]^x[A^{x-}]^y.$$

The values given in the following table are for 298 K unless otherwise stated.

	Compound	$K_{sp}/(\text{mol dm}^{-3})^{x+y}$
Bromides	AgBr	7.7×10^{-13} (291 K)
	$PbBr_2$	7.9×10^{-5} (291 K)
Carbonates	$BaCO_3$	5.1×10^{-9}
	$CaCO_3$	4.8×10^{-9}
	$MgCO_3$	1.0×10^{-5}
	$SrCO_3$	1.1×10^{-10}
Chlorides	AgCl	1.8×10^{-10}
	Hg_2Cl_2	1.3×10^{-18}
	$PbCl_2$	1.6×10^{-5}
Chromates	Ag_2CrO_4	1.3×10^{-12}
	$BaCrO_4$	5.0×10^{-10}
	$PbCrO_4$	1.8×10^{-14} (291 K)
	$SrCrO_4$	3.6×10^{-5} (room temperature)
Hydroxides	$Al(OH)_3$	1.0×10^{-33}
	$Ba(OH)_2$	5.0×10^{-3}
	$Ca(OH)_2$	5.5×10^{-6}
	$Cr(OH)_3$	1.0×10^{-33}
	$Co(OH)_2$	6.3×10^{-15} (blue)
		1.6×10^{-15} (pink-fresh)
		2.0×10^{-16} (pink-aged)
	$Fe(OH)_2$	7.9×10^{-16}
	$Fe(OH)_3$	2.0×10^{-39}
	$Mg(OH)_2$	1.1×10^{-11}
	$Mn(OH)_2$	2.0×10^{-13}
	$Ni(OH)_2$	6.5×10^{-18}
	$Sr(OH)_2$	3.2×10^{-4}
	$Zn(OH)_2$	2.0×10^{-17}
Iodides	AgI	8.3×10^{-17}
	PbI_2	1.0×10^{-9}
Phosphates	Ag_3PO_4	1.4×10^{-21} (286–296 K)
	$MgNH_4PO_4$	2.5×10^{-13} (room temperature)
Sulphates	Ag_2SO_4	1.7×10^{-5}
	$BaSO_4$	1.3×10^{-10}
	$CaSO_4$	2.4×10^{-5}
	$PbSO_4$	1.6×10^{-8}
	$SrSO_4$	3.2×10^{-7}
Sulphides	Ag_2S	6.3×10^{-51}
	Bi_2S_3	1.0×10^{-97}
	CdS	1.6×10^{-28}
	CoS	4.0×10^{-21} (α)
		2.0×10^{-25} (β)
	CuS	6.3×10^{-36}
	FeS	6.3×10^{-18}

	Compound	$K_{sp}/(\text{mol dm}^{-3})^{x+y}$
(Sulphides)	HgS	$1 \cdot 6 \times 10^{-52}$ (black)
		$1 \cdot 4 \times 10^{-53}$ (red)
	MnS	$2 \cdot 5 \times 10^{-10}$ (pink)
		$2 \cdot 5 \times 10^{-13}$ (green)
	NiS	$3 \cdot 2 \times 10^{-19}$ (α)
		$1 \cdot 0 \times 10^{-24}$ (β)
		$2 \cdot 0 \times 10^{-26}$ (γ)
	PbS	$1 \cdot 3 \times 10^{-28}$
	Sb_2S_3	$1 \cdot 7 \times 10^{-93}$
	SnS	$1 \cdot 0 \times 10^{-25}$
	ZnS	$1 \cdot 6 \times 10^{-24}$ (α)
		$2 \cdot 5 \times 10^{-22}$ (β)
Thiocyanates	AgSCN	$1 \cdot 0 \times 10^{-12}$

48 Stability (or formation) Constants of Complex Ions at 298 K

The stability constant of a complex ion is a measure of its stability with respect to dissociation into its constituent species at a given temperature, e.g. the formation of the tetra-amminecopper(II) ion may be represented by the equation

$$Cu^{2+} + 4NH_3 = [Cu(NH_3)_4]^{2+}$$

and the stability constant is given by

$$K_{stab} = \frac{[Cu(NH_3)_4^{2+}]}{[Cu^{2+}][NH_3]^4}.$$

The higher the stability constant the more stable the complex ion. v denotes the stoichiometric number of a molecule, atom or ion, and is positive for a product and negative for a reactant.

(a) General

Equilibrium	$\dfrac{K_{stab}}{(\text{mol dm}^{-3})^{\Sigma v}}$	$\log_{10}\left\{\dfrac{K_{stab}}{(\text{mol dm}^{-3})^{\Sigma v}}\right\}$
$Ag^+ + 2CN^- = [Ag(CN)_2]^-$	$1 \cdot 0 \times 10^{21}$	$21 \cdot 0$
$Ag^+ + NH_3 = [Ag(NH_3)]^+$	$2 \cdot 5 \times 10^3$	$3 \cdot 4$
$[Ag(NH_3)]^+ + NH_3 = [Ag(NH_3)_2]^+$	$6 \cdot 3 \times 10^3$	$3 \cdot 8$
$Ag^+ + 2NH_3 = [Ag(NH_3)_2]^+$	$1 \cdot 7 \times 10^7$	$7 \cdot 2$
$Ag^+ + 2S_2O_3^{2-} = [Ag(S_2O_3)_2]^{3-}$	$1 \cdot 0 \times 10^{13}$	$13 \cdot 0$
$Al^{3+} + 6F^- = [AlF_6]^{3-}$	6×10^{19}	$19 \cdot 8$
$Al(OH)_3 + OH^- = [Al(OH)_4]^-$	40	$1 \cdot 6$
$Cd^{2+} + 4CN^- = [Cd(CN)_4]^{2-}$	$7 \cdot 1 \times 10^{16}$	$16 \cdot 9$
$Cd^{2+} + 4I^- = [CdI_4]^{2-}$	2×10^6	$6 \cdot 3$
$Cd^{2+} + 4NH_3 = [Cd(NH_3)_4]^{2+}$	$4 \cdot 0 \times 10^6$	$6 \cdot 6$
$Co^{2+} + 6NH_3 = [Co(NH_3)_6]^{2+}$	$7 \cdot 7 \times 10^4$	$4 \cdot 9$
$Co^{3+} + 6NH_3 = [Co(NH_3)_6]^{3+}$	$4 \cdot 5 \times 10^{33}$	$33 \cdot 7$

Equilibrium	$\dfrac{K_{stab}}{(mol\ dm^{-3})^{\Sigma v}}$	$\log_{10}\left\{\dfrac{K_{stab}}{(mol\ dm^{-3})^{\Sigma v}}\right\}$
$Cr(OH)_3 + OH^- = [Cr(OH)_4]^-$	$1\ \times 10^{-2}$	-2
$Cu^+ + 4CN^- = [Cu(CN)_4]^{3-}$	$2{\cdot}0 \times 10^{27}$	$27{\cdot}3$
$Cu^{2+} + 4Cl^- = [CuCl_4]^{2-}$	$4{\cdot}0 \times 10^5$	$5{\cdot}6$
$Cu^+ + 2NH_3 = [Cu(NH_3)_2]^+$	$1\ \times 10^{11}$	11
$Cu^{2+} + NH_3 = [Cu(NH_3)]^{2+}$	$2{\cdot}0 \times 10^4\ (K_1)$	$4{\cdot}3$
$[Cu(NH_3)]^{2+} + NH_3 = [Cu(NH_3)_2]^{2+}$	$4{\cdot}2 \times 10^3\ (K_2)$	$3{\cdot}6$
$[Cu(NH_3)_2]^{2+} + NH_3 = [Cu(NH_3)_3]^{2+}$	$1{\cdot}0 \times 10^3\ (K_3)$	$3{\cdot}0$
$[Cu(NH_3)_3]^{2+} + NH_3 = [Cu(NH_3)_4]^{2+}$	$1{\cdot}7 \times 10^2\ (K_4)$	$2{\cdot}2$
$Cu^{2+} + 4NH_3 = [Cu(NH_3)_4]^{2+}$	$1{\cdot}4 \times 10^{13}$	$13{\cdot}1$
	$(K = K_1 K_2 K_3 K_4)$	
$Fe^{2+} + 6CN^- = [Fe(CN)_6]^{4-}$	ca. 10^{24}	ca. 24
$Fe^{3+} + 6CN^- = [Fe(CN)_6]^{3-}$	ca. 10^{31}	ca. 31
$Fe^{3+} + 4Cl^- = [FeCl_4]^-$	$8\ \times 10^{-2}$	$-1{\cdot}1$
$Fe^{3+} + SCN^- = [Fe(SCN)]^{2+}$	$1{\cdot}4 \times 10^2$	$2{\cdot}1$
$[Fe(SCN)]^{2+} + SCN^- = [Fe(SCN)_2]^+$	16	$1{\cdot}2$
$[Fe(SCN)_2]^+ + SCN^- = Fe(SCN)_3$	1	0
$Hg^{2+} + 4CN^- = [Hg(CN)_4]^{2-}$	$2{\cdot}5 \times 10^{41}$	$41{\cdot}4$
$Hg^{2+} + 4Cl^- = [HgCl_4]^{2-}$	$1{\cdot}7 \times 10^{16}$	$16{\cdot}2$
$Hg^{2+} + 4I^- = [HgI_4]^{2-}$	$2{\cdot}0 \times 10^{30}$	$30{\cdot}3$
$I^- + I_2 = I_3^-$	$7{\cdot}1 \times 10^2$	$2{\cdot}9$
$Ni^{2+} + 6NH_3 = [Ni(NH_3)_6]^{2+}$	$4{\cdot}8 \times 10^7$	$7{\cdot}7$
$Pb(OH)_2 + OH^- = [Pb(OH)_3]^-$	50	$1{\cdot}7$
$Sn(OH)_4 + 2OH^- = [Sn(OH)_6]^{2-}$	$5\ \times 10^3$	$3{\cdot}7$
$Zn^{2+} + 4CN^- = [Zn(CN)_4]^{2-}$	$5\ \times 10^{16}$	$16{\cdot}7$
$Zn^{2+} + 4NH_3 = [Zn(NH_3)_4]^{2+}$	$3{\cdot}8 \times 10^9$	$9{\cdot}6$
$Zn(OH)_2 + 2OH^- = [Zn(OH)_4]^{2-}$	10	$1{\cdot}0$

(b) Metal-edta Complexes

Cation	$\log_{10}(K_{stab}/mol^{-1}\ dm^3)$	Cation	$\log_{10}(K_{stab}/mol^{-1}\ dm^3)$
Ag^+	$7{\cdot}3$	Fe^{2+}	$14{\cdot}3$
Al^{3+}	$16{\cdot}1$	Fe^{3+}	$25{\cdot}1$
Ba^{2+}	$7{\cdot}8$	Hg^{2+}	$21{\cdot}8$
Ca^{2+}	$10{\cdot}7$	Li^+	$2{\cdot}8$
Cd^{2+}	$16{\cdot}6$	Mg^{2+}	$8{\cdot}7$
Co^{2+}	$16{\cdot}3$	Mn^{2+}	$14{\cdot}0$
Co^{3+}	36	Na^+	$1{\cdot}7$
Cr^{2+}	$13{\cdot}0$	Ni^{2+}	$18{\cdot}6$
Cr^{3+}	$24{\cdot}0$	Pb^{2+}	$18{\cdot}0$
Cu^{2+}	$18{\cdot}8$	Sr^{2+}	$8{\cdot}6$
		Zn^{2+}	$16{\cdot}5$

Note. The formula of edta (ethylenediamine tetra-acetic acid) is

$$\begin{array}{ccc}
HO_2C.H_2C & & CH_2.CO_2H \\
& >\!N.CH_2.CH_2.N\!< & \\
HO_2C.H_2C & & CH_2.CO_2H
\end{array}$$

Its systematic name is bis[di(carboxymethyl)amino]ethane. It is usually written as H_4Y, so that the formation of a metal-edta complex may be represented by the equation

$$M^{z+} + Y^{4-} = MY^{z-4}$$

and the stability constant is given by

$$K_{stab} = \frac{[MY^{z-4}]}{[M^{z+}][Y^{4-}]}$$

49 Electrolytic Conductivity of Potassium Chloride Solution

The values of electrolytic conductivity κ in the following table may be used for calibrating conductivity cells: they are given for various concentrations c of the solute at a temperature θ_C. Siemens $S = \Omega^{-1}$.

| | $\theta_C/°C$ | $c/mol\ dm^{-3}$ | | |
		0.001	0.01	0.10
$\kappa/S\ m^{-1}$	18	0.0127	0.122	1.12
	25	0.0147	0.141	1.29

50 Molar Conductivity of Aqueous Solutions at 298 K

The values of molar conductivity Λ in the following table are given for various concentrations c of the solute. Siemens $S = \Omega^{-1}$.

Compound		$10^4 \Lambda_+ /S\ m^2\ mol^{-1}$		
$c/mol\ dm^{-3}$	0	0·001	0·01	0·10
$AgNO_3$	133·4	130·5	124·8	109·1
HCl	426·2	421·4	412·0	391·3
CH_3CO_2H	390·7	48·7	16·2	5·0
KBr	151·9	149·0	143·4	131·4
KCl	149·9	147·0	141·3	129·0
KI	150·4	147·3	142·2	137·1
KNO_3	145·0	141·8	132·8	130·4
LiCl	115·0	112·4	107·3	95·9
NH_4Cl	149·7	146·8	141·3	128·8
NaCl	126·5	123·7	118·5	106·7
NaI	126·9	124·2	119·2	108·8
NaOH	248·4	244·7	238·0	218·2
CH_3CO_2Na	91·0	88·5	83·8	72·8

51 Molar Conductivity of Ions at Infinite Dilution at 298 K

Λ denotes molar conductivity. Siemens $S = \Omega^{-1}$.

Cation	$10^4 \Lambda_+ /S\ m^2\ mol^{-1}$	Anion	$10^4 \Lambda_- /S\ m^2\ mol^{-1}$
H^+	349·8	OH^-	198·6
Li^+	38·7	F^-	55·4
Na^+	50·1	Cl^-	76·4
K^+	73·5	Br^-	78·1
NH_4^+	73·5	I^-	76·8
Ag^+	61·9	CO_3^{2-}	118·6
Mg^{2+}	106·1	HCO_3^-	44·5
Ca^{2+}	119·0	CN^-	82
Sr^{2+}	118·9	NO_3^-	71·4
Ba^{2+}	127·2	PO_4^{3-}	240
Zn^{2+}	105·6	SO_4^{2-}	159·6
Hg^{2+}	127·2	HSO_4^-	52
Al^{3+}	189	ClO_3^-	64·6
Pb^{2+}	118·9	BrO_3^-	55·8
Cu^{2+}	107·2	IO_3^-	40·5
Mn^{2+}	107·0	ClO_4^-	67·4
Fe^{2+}	108	MnO_4^-	61
Fe^{3+}	205·2	HCO_2^-	54·6
Co^{2+}	110	$CH_3CO_2^-$	40·9
Ni^{2+}	106	$C_2O_4^{2-}$	148·4

52 Electrolytic Conductivity of Water at Various Temperatures

κ denotes electrolytic conductivity. Siemens $S = \Omega^{-1}$.
θ_C denotes Celsius temperature.

$\theta_C/°C$	$10^6\kappa/S\,m^{-1}$
0	1·2
10	2·3
20	4·2
25	5·5
30	7·1
40	11·3
50	17·1

53 Kinetic Data

t denotes time and c concentration.

(a) Hydrolysis of Methyl Ethanoate

Methyl ethanoate is hydrolysed in aqueous solution in the presence of an acid catalyst as follows:

$$CH_3CO_2CH_3 + H_2O = CH_3CO_2H + CH_3OH$$

The course of the reaction can be followed titrimetrically.

t/s	$c(CH_3CO_2CH_3)/mol\ dm^{-3}$
0	0·30
1800	0·19
3600	0·13
5400	0·082
7200	0·053
9000	0·035

(b) Thermal Decomposition of Dinitrogen Pentoxide

The decomposition of dinitrogen pentoxide in the gas phase

$$N_2O_5(g) = N_2O_4(g) + \tfrac{1}{2}O_2(g)$$

involves an increase in pressure. The course of the reaction can, therefore, be followed by measurement of the total pressure p of the system. The data in the following table relate to a decomposition temperature of 308 K.

t/s	$p/kN\ m^{-2}$
0	41·1
1200	44·7
2400	47·1
3600	49·2
6000	52·5
8400	54·9
12000	57·4
∞	61·6

(c) Inversion of Cane Sugar

The hydrolysis of cane sugar (sucrose) to glucose and fructose in the presence of dilute hydrochloric acid

$$C_{12}H_{22}O_{11} + H_2O = C_6H_{12}O_6 + C_6H_{12}O_6$$

sucrose   glucose   fructose
(dextrorotatory)   (laevorotatory)

leads to a reversal of the optical rotation (inversion). The course of the reaction can, therefore, be followed by polarimetric measurement of the angle α of optical rotation. The following data relate to a reaction temperature of 298 K.

t/s	$\alpha/°$
0	$+24.09$
430.8	$+21.40$
1080	$+17.74$
1623	$+15.00$
2208	$+12.40$
2760	$+10.02$
3364	$+7.80$
4081	$+5.46$
6102	$+0.30$
∞	-10.74

(d) Saponification of Ethyl Ethanoate

Ethyl ethanoate undergoes hydrolysis in alkaline solution:

$$CH_3CO_2C_2H_5 + OH^- = CH_3CO_2^- + C_2H_5OH$$

The course of the reaction can be followed titrimetrically. The following data relate to a reaction temperature of 289 K.

Initial concentration of $CH_3CO_2C_2H_5$ = initial concentration of OH^- = 0.06400 mol dm^{-3}.

t/s	$c(CH_3CO_2^-)$/mol dm^{-3}
0	0.00000
300	0.02304
900	0.03948
1500	0.04672
2100	0.05036
3300	0.05472
∞	0.06400

(e) Effect of Temperature on Rate of Reaction

The activation energy E for a reaction can be calculated from the Arrhenius equation

$$k = A\, e^{-E/RT}$$

where k denotes the rate constant, R denotes the gas constant, T denotes the temperature and A is called the frequency factor.

$H_2 + I_2 = 2HI$

T/K	556	575	629	666	700	781
$k/\text{dm}^3\,\text{mol}^{-1}\,\text{s}^{-1}$	4.45×10^{-5}	1.37×10^{-4}	2.52×10^{-3}	1.41×10^{-2}	6.43×10^{-2}	1.34

$2HI = H_2 + I_2$

T/K	556	575	629	666	700	781
$k/\text{dm}^3\,\text{mol}^{-1}\,\text{s}^{-1}$	7.04×10^{-7}	2.44×10^{-6}	6.04×10^{-5}	4.38×10^{-4}	2.32×10^{-3}	7.90×10^{-2}

$2N_2O_5 = 2N_2O_4 + O_2$

T/K	273	293	298	308	318	328	338
k/s^{-1}	7.87×10^{-7}	1.76×10^{-5}	3.38×10^{-5}	1.35×10^{-4}	4.98×10^{-4}	1.50×10^{-3}	4.87×10^{-3}

$2N_2O = 2N_2 + O_2$

T/K	838	1001	1030	1053	1125
k/s^{-1}	1.10×10^{-3}	3.80×10^{-1}	8.71×10^{-1}	1.67	11.6

54 Experimental Rate Laws

If a chemical reaction can be represented by the *mechanistic* equation

$$A + B = products$$

then, according to the kinetic law of mass action, the rate equation is $-d[A]/dt = -d[B]dt = k\,[A]\,[B]$ where k denotes the rate constant for the reaction. The sum of the powers to which the concentrations of the reactants are raised in the rate expression is known as the overall *order* of the reaction.

Reaction	Rate equation
$H_2 + I_2 = 2HI$	$-d[I_2]/dt = k[H_2]\,[I_2]$
$H_2 + Br_2 = 2HBr$	$-d[Br_2]/dt = k[H_2]\,[Br_2]^{3/2}/([Br_2] + k'[HBr])$
$2NO + O_2 = 2NO_2$	$-d[O_2]/dt = -\frac{1}{2}d[NO]/dt = k[NO]^2[O_2]$
$2N_2O_5 = 4NO_2 + O_2$	$-d[N_2O_5]/dt = k[N_2O_5]$
$H_2O_2 + 2H^+ + 2I^- = 2H_2O + I_2$	$-d[H_2O_2]/dt = k[H_2O_2]\,[I^-]\,(1 + k'\,[H^+])$
$S_2O_8^{2-} + 2I^- = 2SO_4^{2-} + I_2$	$-d[S_2O_8^{2-}]/dt = k[S_2O_8^{2-}]\,[I^-]$
$BrO_3^- + 5Br^- + 6H^+ = 3Br_2 + 3H_2O$	$-d[BrO_3^-]/dt = k[BrO_3^-]\,[Br^-]\,[H^+]^2$
$CH_3Br + OH^- = CH_3OH + Br^-$	$-d[CH_3Br]/dt = k[CH_3Br]\,[OH^-]$
$(CH_3)_3C.Cl + H_2O = (CH_3)_3C.OH + Cl^-$	$-d[(CH_3)_3C.Cl]/dt = k[(CH_3)_3C.Cl]$
$CH_3CO_2H + C_2H_5OH = CH_3CO_2C_2H_5 + H_2O$	$-d[CH_3CO_2H]/dt = k[CH_3CO_2H]\,[C_2H_5OH]$
$CH_3CO_2C_2H_5 + OH^- = CH_3CO_2^- + C_2H_5OH$	$-d[CH_3CO_2C_2H_5]/dt = k[CH_3CO_2C_2H_5]\,[OH^-]$
$I_2 + CH_3COCH_3 \overset{H^+}{=} CH_2I.CO.CH_3 + HI$	$-d[I_2]/dt = k[CH_3COCH_3]\,[H^+]$

55 Activation Energies

According to the Arrhenius equation, the rate constant k of a chemical reaction is related to the temperature T as follows:

$$k = A\, e^{-E/RT}$$

where E is known as the activation energy.

Reaction	Catalyst	$E/\text{kJ mol}^{-1}$
$2HI = H_2 + I_2$	None	183
	Au	105
	Pt	58
$H_2 + I_2 = 2HI$	None	157
$H_2 + Cl_2 = 2HCl$ (photochemical)	None	25
$2NH_3 = N_2 + 3H_2$	None	330 approx.
	W	163
$2N_2O = 2N_2 + O_2$	None	245
	Au	121
	Pt	136
$2NO_2 = 2NO + O_2$	None	112
$2NOCl = 2NO + Cl_2$	None	98·7
$2NOBr = 2NO + Br_2$	None	58·2
$2H_2O_2 = 2H_2O + O_2$	None	75·3
	Colloidal Pt	48·9
	Enzyme catalase	23
$CH_3CHO = CH_4 + CO$	None	190
	I_2 vapour	136
$C_2H_5OC_2H_5 = C_2H_6 + CO + CH_4$	None	224
	I_2 vapour	143

56 Physical Properties of Organic Compounds

$\theta_{C,m}$ denotes melting temperature
$\theta_{C,b}$ denotes boiling temperature
ρ denotes density of substance at 20 °C
n_D denotes refractive index for the
   D line of sodium at 20 °C
solubility denotes solubility in water at 20 °C

dec. denotes decomposes
subl. denotes sublimes
expl. denotes explodes
vs denotes very soluble
s denotes soluble
ss denotes slightly soluble
i denotes insoluble

Name and formula	$\theta_{C.m}/°C$	$\theta_{C.b}/°C$	$\rho/\text{g cm}^{-3}$	n_D	Solu-bility
ALKANES					
Methane, CH_4	−182	−162			i
Ethane, CH_3CH_3	−183	−88·6			i
Propane, $CH_3CH_2CH_3$	−188	−42·2			i
Butane, $CH_3(CH_2)_2CH_3$	−138	−0·5			i
Pentane, $CH_3(CH_2)_3CH_3$	−130	36·3	0·626	1·358	i
Hexane, $CH_3(CH_2)_4CH_3$	−95·3	68·7	0·659	1·375	i
Hexadecane, $CH_3(CH_2)_{14}CH_3$	18·5	287	0·775	1·434	i
2-Methylpropane, $(CH_3)_3CH$	−160	−11·7			i
2-Methylbutane, $(CH_3)_2CHCH_2CH_3$	−158	27·9	0·620	1·354	i
2,2-Dimethylpropane, $(CH_3)_4C$	−15·9	9·5			
CYCLOALKANES					
Cyclopropane, $CH_2CH_2CH_2$	−127	−32·9			i
Cyclobutane, $CH_2CH_2CH_2CH_2$	−50	13			i
Cyclopentane, $CH_2CH_2CH_2CH_2CH_2$	−93·9	49·3	0·751	1·406	i
Cyclohexane, $CH_2CH_2CH_2CH_2CH_2CH_2$	6·5	80·7	0·779	1·426	i
ALKENES					
Ethene, $CH_2=CH_2$	−169	−104			i
Propene, $CH_3CH=CH_2$	−185	−47·7			i
But-1-ene, $CH_3CH_2CH=CH_2$	−185	−6·2			i
Pent-1-ene, $CH_3(CH_2)_2CH=CH_2$	−138	30·0	0·640	1·372	i
Hex-1-ene, $CH_3(CH_2)_3CH=CH_2$	−98	63·9	0·674	1·388	i
cis-But-2-ene, $CH_3CH=CHCH_3$	−139	3·7			i
trans-But-2-ene, $CH_3CH=CHCH_3$	−106	0·9			i
2-Methylpropene, $CH_3C(CH_3)=CH_2$	−139	−6·6			i
Cyclohexene, $CH_2(CH_2)_3CH=CH$	−104	83·3	0·810	1·445	i
Buta-1,3-diene, $CH_2=CHCH=CH_2$	−109	−4·4			
ALKYNES					
Ethyne, $CH\equiv CH$	subl.	−83·6			i
Propyne, $CH_3C\equiv CH$	−103	−23·2			i
But-1-yne, $CH_3CH_2C\equiv CH$	−122	8·1			i

Name and formula	$\theta_{C,m}/°C$	$\theta_{C,b}/°C$	$\rho/g\,cm^{-3}$	n_D	Solu-bility
But-2-yne, $CH_3C\equiv CCH_3$	−32·3	27·0	0·691	1·392	i
Pent-1-yne, $CH_3(CH_2)_2C\equiv CH$	−90·0	39·3	0·695	1·385	i
Hex-1-yne, $CH_3(CH_2)_3C\equiv CH$	−132	71	0·716	1·398	i
AROMATIC HYDROCARBONS					
Benzene, C_6H_6	5·5	80·1	0·878	1·501	i
Methylbenzene, $C_6H_5CH_3$	−95·0	111	0·867	1·497	i
Ethylbenzene, $C_6H_5CH_2CH_3$	−94	136	0·867	1·496	i
Propylbenzene, $C_6H_5(CH_2)_2CH_3$	−99·5	159	0·862	1·493	i
(1-Methylethyl)benzene, $C_6H_5CH(CH_3)_2$	−96	152	0·864	1·491	i
1,2-Dimethylbenzene, $C_6H_4(CH_3)_2$	−25·2	144	0·880	1·505	i
1,3-Dimethylbenzene, $C_6H_4(CH_3)_2$	−47·4	139	0·864	1·497	i
1,4-Dimethylbenzene, $C_6H_4(CH_3)_2$	13·3	138	0·861	1·496	i
Naphthalene, $C_{10}H_8$	80·2	218	1·145		i
Anthracene, $C_{14}H_{10}$	216	340	1·243		i
HALOGEN COMPOUNDS					
Chloro-derivatives:					
Chloromethane, CH_3Cl	−97·7	−23·8			ss
Chloroethane, CH_3CH_2Cl	−136	12·5			ss
1-Chloropropane, $CH_3(CH_2)_2Cl$	−123	46·6	0·889	1·388	ss
1-Chlorobutane, $CH_3(CH_2)_3Cl$	−123	78·5	0·886	1·402	i
1-Chloropentane, $CH_3(CH_2)_4Cl$	−99	108	0·883	1·412	i
1-Chlorohexane, $CH_3(CH_2)_5Cl$	−83	133	0·878	1·420	i
(Chloromethyl)benzene, $C_6H_5CH_2Cl$	−39	179	1·100	1·539	i
Chlorobenzene, C_6H_5Cl	−45·2	132	1·106	1·525	i
2-Chloropropane, $(CH_3)_2CHCl$	−117	34·8	0·859	1·378	ss
1-Chloro-2-methylpropane, $(CH_3)_2CHCH_2Cl$	−131	68	0·881	1·398	i
2-Chlorobutane, $CH_3CH_2CHClCH_3$	−131	67	0·874	1·397	i
2-Chloro-2-methylpropane, $(CH_3)_3CCl$	−27·1	50·7	0·851	1·390	ss
Dichloromethane, CH_2Cl_2	−96·8	40·2	1·336	1·424	ss
1,1-Dichloroethane, CH_3CHCl_2	−97·4	57·3	1·178	1·417	i
1,2-Dichloroethane, CH_2ClCH_2Cl	−35·3	83·7	1·253	1·444	i
Trichloromethane, $CHCl_3$	−63·5	61·2	1·489	1·447	i
Tetrachloromethane, CCl_4	−23·0	76·8	1·594	1·461	i
Difluorodichloromethane, CF_2Cl_2	−158	−29·8			i
1,2-Dichlorobenzene, $C_6H_4Cl_2$	−17	179	1·305	1·549	i
1,3-Dichlorobenzene, $C_6H_4Cl_2$	−24	172	1·288	1·546	i
1,4-Dichlorobenzene, $C_6H_4Cl_2$	53	174			i
Chloroethene, $CH_2=CHCl$	−160	−13·9			ss
Bromo-derivatives:					
Bromomethane, CH_3Br	−93·7	3·6			ss
Bromoethane, CH_3CH_2Br	−119	38·4	1·460	1·424	ss
1-Bromopropane, $CH_3(CH_2)_2Br$	−109	70·8	1·354	1·434	ss
1-Bromobutane, $CH_3(CH_2)_3Br$	−113	101	1·279	1·440	i
1-Bromopentane, $CH_3(CH_2)_4Br$	−95	129	1·218	1·444	i
1-Bromohexane, $CH_3(CH_2)_5Br$	−85	156	1·176	1·448	i
(Bromomethyl)benzene, $C_6H_5CH_2Br$	−3·9	201	1·438	1·575	i
Bromobenzene, C_6H_5Br	−30·6	156	1·494	1·560	i

Name and formula	$\theta_{C,m}/°C$	$\theta_{C,b}/°C$	$\rho/g\ cm^{-3}$	n_D	Solubility
2-Bromopropane, $(CH_3)_2CHBr$	−90·8	59·4	1·310	1·425	ss
1-Bromo-2-methylpropane, $(CH_3)_2CHCH_2Br$		91	1·253	1·435	i
2-Bromobutane, $CH_3CH_2CHBrCH_3$	−112	91·2	1·256	1·437	i
2-Bromo-2-methylpropane, $(CH_3)_3CBr$	−20	73·3	1·222	1·428	i
Dibromomethane, CH_2Br_2	−52·7	97·0	2·495	1·542	ss
1,1-Dibromoethane, CH_3CHBr_2	−63	110	2·055	1·512	i
1,2-Dibromoethane, CH_2BrCH_2Br	10·1	132	2·180	1·539	i
Tribromomethane, $CHBr_3$	8·3	150	2·890	1·598	i
Tetrabromomethane, CBr_4	92	190			i
1,2-Dibromobenzene, $C_6H_4Br_2$	6·7	221	1·956	1·610	i
1,3-Dibromobenzene, $C_6H_4Br_2$	−7	220	1·952	1·608	i
1,4-Dibromobenzene, $C_6H_4Br_2$	87	218			i
Bromoethene, $CH_2=CHBr$	−138	16	1·517	1·446	i
Iodo-derivatives:					
Iodomethane, CH_3I	−66·5	42·5	2·279	1·532	ss
Iodoethane, CH_3CH_2I	−108	72·4	1·940	1·513	ss
1-Iodopropane, $CH_3(CH_2)_2I$	−101	102	1·745	1·506	i
1-Iodobutane, $CH_3(CH_2)_3I$	−103	130	1·617	1·500	i
1-Iodopentane, $CH_3(CH_2)_4I$	−85·6	155	1·517	1·496	i
1-Iodohexane, $CH_3(CH_2)_5I$		181	1·437	1·493	i
(Iodomethyl)benzene, $C_6H_5CH_2I$	24·5	dec.	1·734	1·633	i
Iodobenzene, C_6H_5I	−31·4	189	1·831	1·620	i
2-Iodopropane, $(CH_3)_2CHI$	−90·1	89·4	1·704	1·500	ss
1-Iodo-2-methylpropane, $(CH_3)_2CHCH_2I$		119	1·602	1·496	i
2-Iodobutane, $CH_3CH_2CHICH_3$	−104	118	1·598	1·499	i
2-Iodo-2-methylpropane, $(CH_3)_3CI$	−38·2	98	1·545	1·492	dec.
Diiodomethane, CH_2I_2	6·1	180 dec.	3·325	1·745	ss
1,1-Diiodoethane, CH_3CHI_2		179–180			i
1,2-Diiodoethane, CH_2ICH_2I	81–82				i
Triiodomethane, CHI_3	119	subl.			i
Tetraiodomethane, CI_4	subl.	dec.			i
1,2-Diiodobenzene, $C_6H_4I_2$	27	286			i
1,3-Diiodobenzene, $C_6H_4I_2$	40·4	285			i
1,4-Diiodobenzene, $C_6H_4I_2$	129	285 subl.			i
Iodoethene, $CH_2=CHI$		56			i
ALCOHOLS					
Methanol, CH_3OH	−97·7	64·5	0·791	1·329	vs
Ethanol, CH_3CH_2OH	−117	78·5	0·789	1·361	vs
Propan-1-ol, $CH_3(CH_2)_2OH$	−127	97·2	0·803	1·385	vs
Butan-1-ol, $CH_3(CH_2)_3OH$	−89·5	117	0·810	1·399	s
Pentan-1-ol, $CH_3(CH_2)_4OH$	−78·8	138	0·814	1·410	ss
Hexan-1-ol, $CH_3(CH_2)_5OH$	−51·6	157	0·814	1·418	ss
Phenylmethanol, $C_6H_5CH_2OH$	−15·3	205	1·042	1·540	ss
Propan-2-ol, $(CH_3)_2CHOH$	−89·5	82·4	0·785	1·378	vs
2-Methylpropan-1-ol, $(CH_3)_2CHCH_2OH$	−108	108	0·802	1·396	s
Butan-2-ol, $CH_3CH_2CHOHCH_3$	−115	99·5	0·808	1·393	s

Name and formula	$\theta_{C,m}/^{\circ}C$	$\theta_{C,b}/^{\circ}C$	$\rho/g\ cm^{-3}$	n_D	Solu-bility
2-Methylpropan-2-ol, $(CH_3)_3COH$	25·5	82·5	0·786	1·388	vs
Cyclohexanol, $\underline{CH_2(CH_2)_4}CHOH$	25·2	161	0·962	1·463	ss
Ethane-1,2-diol, CH_2OHCH_2OH	−15·6	198	1·109	1·427	vs
Propane-1,2,3-triol, $CH_2OHCHOHCH_2OH$	18·2	290 dec.	1·261	1·475	vs
PHENOLS					
Phenol, C_6H_5OH	40·9	182	1·073	1·551	{s(hot) ss
2-Methylphenol, $CH_3C_6H_4OH$	31·0	191	1·046	1·445	{vs(hot) s
3-Methylphenol, $CH_3C_6H_4OH$	12·0	202	1·034	1·540	{s(hot) ss
4-Methylphenol, $CH_3C_6H_4OH$	34·7	202	1·035	1·540	{s(hot) ss
Benzene-1,2-diol, $C_6H_4(OH)_2$	105	240	1·371	1·604	s
Benzene-1,3-diol, $C_6H_4(OH)_2$	111	280			s
Benzene-1,4-diol, $C_6H_4(OH)_2$	170	286			{vs(hot) s
2-Nitrophenol, $NO_2C_6H_4OH$	44·9	216			ss(hot)
3-Nitrophenol, $NO_2C_6H_4OH$	97	dec.			{vs(hot) ss
4-Nitrophenol, $NO_2C_6H_4OH$	114	279 dec.			{vs(hot) ss
2-Chlorophenol, ClC_6H_4OH	8·7	175–176	1·241	1·558	vs
3-Chlorophenol, ClC_6H_4OH	32·8	213–216			{s(hot) ss
4-Chlorophenol, ClC_6H_4OH	43	220			s
Naphthalen-1-ol, $C_{10}H_7OH$	93·4	288			{ss(hot) i
Naphthalen-2-ol, $C_{10}H_7OH$	122	295			i
2.4.6-Trinitrophenol, $(NO_2)_3C_6H_2OH$	122	expl.	1·763		{s(hot) ss
ETHERS					
Methoxymethane, CH_3OCH_3	−142	−24·8			s
Methoxyethane, $CH_3OCH_2CH_3$		7·0			s
Ethoxyethane, $CH_3CH_2OCH_2CH_3$	−116	34·5	0·714	1·353	ss
Methoxypropane, $CH_3O(CH_2)_2CH_3$		38·9	0·738	1·358	ss
Methoxy-1-methylethane, $CH_3OCH(CH_3)_2$		32·5	0·735	1·358	ss
Methoxybutane, $CH_3O(CH_2)_3CH_3$	−116	70·3	0·744	1·374	i
Methoxybenzene (anisole), $CH_3OC_6H_5$	−37·5	154	0·995	1·518	i
Ethoxybenzene (phenetole), $CH_3CH_2OC_6H_5$	−30	172	0·970	1·509	i
Phenoxybenzene, $C_6H_5OC_6H_5$	28	258			i
Epoxyethane, $\underline{CH_2CH_2O}$	−111	10·7			s
ALDEHYDES					
Methanal, $HCHO$	−92	−21			vs
Ethanal, CH_3CHO	−124	20·8	0·783	1·332	vs
Propanal, CH_3CH_2CHO	−81	48·8	0·807	1·364	s
Butanal, $CH_3(CH_2)_2CHO$	−99	75·7	0·817	1·382	ss

Name and formula	$\theta_{C,m}/°C$	$\theta_{C,b}/°C$	$\rho/g\ cm^{-3}$	n_D	Solu-bility
2-Methylpropanal, $(CH_3)_2CHCHO$	-65.9	64.2	0.790	1.372	s
Benzaldehyde, C_6H_5CHO	-26	179	1.046	1.546	i
Ethanedial, $CHOCHO$	15	50.4	1.14	1.363	vs
Trichloroethanal, CCl_3CHO	-57.5	97.8	1.512	1.456	vs
KETONES					
Propanone, CH_3COCH_3	-95.4	56.2	0.791	1.359	vs
Butanone, $CH_3COCH_2CH_3$	-86.9	79.6	0.806	1.379	s
Pentan-3-one, $CH_3CH_2COCH_2CH_3$	-39.9	102	0.814	1.392	ss
Pentan-2-one, $CH_3CO(CH_2)_2CH_3$	-77.8	102	0.811	1.390	ss
3-Methylbutan-2-one,$CH_3COCH(CH_3)_2$	-92	95	0.803	1.389	ss
Hexan-2-one, $CH_3CO(CH_2)_3CH_3$	-56.9	127	0.812	1.402	ss
Phenylethanone, $CH_3COC_6H_5$	19.6	202	1.028	1.534	i
Diphenylmethanone, $C_6H_5COC_6H_5$	48.1	306			i
Cyclohexanone, $\underline{CH_2(CH_2)_4C}O$	-16.4	156	0.998	1.452	ss
CARBOXYLIC ACIDS					
Methanoic, HCO_2H	8.4	101	1.220	1.371	vs
Ethanoic, CH_3CO_2H	16.6	118	1.049	1.372	vs
Propanoic, $CH_3CH_2CO_2H$	-20.8	141	0.992	1.387	vs
Butanoic, $CH_3(CH_2)_2CO_2H$	-6.5	164	0.964	1.399	s
Pentanoic, $CH_3(CH_2)_3CO_2H$	-34.5	186	0.939	1.409	s
Hexanoic, $CH_3(CH_2)_4CO_2H$	-1.5	205	0.927	1.416	ss
Phenylethanoic, $C_6H_5CH_2CO_2H$	76	266			s
Benzoic, $C_6H_5CO_2H$	122	249			i
2-Methylpropanoic, $(CH_3)_2CHCO_2H$	-47	154	0.950	1.393	s
Chloroethanoic, $ClCH_2CO_2H$	63	189			vs
Dichloroethanoic, Cl_2CHCO_2H	10.8	194	1.566		vs
Trichloroethanoic, Cl_3CCO_2H	57.5	198			vs
Hydroxyethanoic (glycollic), $HOCH_2CO_2H$	79	dec.			vs
Aminoethanoic (glycine), $H_2NCH_2CO_2H$	232 dec.	286 dec.			vs
2-Hydroxypropanoic (lactic), $CH_3CHOHCO_2H$	16.8	dec.	1.249		vs
cis-Butenedioic (maleic), $HO_2CCH = CHCO_2H$	130–131	dec.	1.590		vs
trans-Butenedioic (fumaric) $HO_2CCH = CHCO_2H$	286 dec.	dec.	1.635		s
Ethanedioic, HO_2CCO_2H	190 dec.				ss
Propanedioic, $HO_2CCH_2CO_2H$	136 dec.	dec.			ss
Butanedioic, $HO_2C(CH_2)_2CO_2H$	182	235 dec.			ss
Benzene-1,2-dicarboxylic, $C_6H_4(CO_2H)_2$	206 dec.	dec.			i
SULPHONIC ACIDS					
Benzenesulphonic acid, $C_6H_5SO_3H$	50–51				s
2-Methylbenzenesulphonic acid, $CH_3C_6H_4SO_3H$	67.5	128			vs

Name and formula	$\theta_{C,m}/$ C	$\theta_{C,b}/°C$	$\rho/g\ cm^{-3}$	n_D	Solubility
4-Methylbenzenesulphonic acid, $CH_3C_6H_4SO_3H$	104	140			vs
4-Aminobenzenesulphonic acid, $NH_2C_6H_4SO_3H$	288				s(hot)
ACID CHLORIDES					
Ethanoyl, CH_3COCl	−112	51	1·105	1·390	dec.
Propanoyl, CH_3CH_2COCl	−94	80	1·065	1·405	dec.
Butanoyl, $CH_3(CH_2)_2COCl$	−89	102	1·028	1·412	dec.
Benzoyl, C_6H_5COCl	−0·5	197	1·210	1·554	dec.
4-Methylbenzenesulphonyl, $CH_3C_6H_4SO_2Cl$	71				i
ACID ANHYDRIDES					
Ethanoic, $(CH_3CO)_2O$	−73·1	140	1·082	1·390	dec.
Propanoic, $(CH_3CH_2CO)_2O$	−45	168	1·022	1·404	dec.
Butanoic, $(CH_3(CH_2)_2CO)_2O$	−75	198	0·967	1·412	dec.
Benzoic, $(C_6H_5CO)_2O$	42	360			i
cis-Butenedioic, $\underset{O}{OCCH = CHCO}$	56	198			s
Butanedioic, $(CH_2CO)_2O$	120	261			ss
Benzene-1,2-dicarboxylic, $C_6H_4(CO)_2O$	131	284 subl.			ss
ESTERS					
Methyl methanoate, HCO_2CH_3	−99	31·5	0·974	1·343	vs
Ethyl methanoate, $HCO_2CH_2CH_3$	−80·5	54·3	0·912	1·360	s
Propyl methanoate, $HCO_2(CH_2)_2CH_3$	−92·9	81·3	0·901	1·377	ss
1-Methylethyl methanoate, $HCO_2CH(CH_3)_2$		70	0·873		ss
Butyl methanoate, $HCO_2(CH_2)_3CH_3$	−91·9	107	0·892	1·389	ss
Phenyl methanoate, $HCO_2C_6H_5$		173			i
Phenylmethyl methanoate, $HCO_2CH_2C_6H_5$		203	1·081	1·575	i
Methyl ethanoate, $CH_3CO_2CH_3$	−98·1	57·3	0·933	1·359	s
Ethyl ethanoate, $CH_3CO_2CH_2CH_3$	−83·6	77·1	0·901	1·372	ss
Propyl ethanoate, $CH_3CO_2(CH_2)_2CH_3$	−95	102	0·888	1·385	ss
1-Methylethyl ethanoate, $CH_3CO_2CH(CH_3)_2$	−73·4	93	0·872	1·377	ss
Butyl ethanoate, $CH_3CO_2(CH_2)_3CH_3$	−77·9	126	0·881	1·394	ss
Phenyl ethanoate, $CH_3CO_2C_6H_5$		196	1·078	1·509	i
Phenylmethyl ethanoate, $CH_3CO_2CH_2C_6H_5$	−51·5	216	1·056	1·503	i
Methyl propanoate, $CH_3CH_2CO_2CH_3$	−87·5	78·7	0·915	1·378	ss
Ethyl propanoate, $CH_3CH_2CO_2CH_2CH_3$	−73·9	99·1	0·889	1·384	ss
Propyl propanoate, $CH_3CH_2CO_2(CH_2)_2CH_3$	−75·9	123	0·881	1·394	ss
1-Methylethyl propanoate, $CH_3CH_2CO_2CH(CH_3)_2$		111			ss

Name and formula	$\theta_{C,m}/°C$	$\theta_{C,b}/°C$	$\rho/g\ cm^{-3}$	n_D	Solubility
Butyl propanoate, $CH_3CH_2CO_2(CH_2)_3CH_3$	-89.6	146	0.875	1.401	ss
Phenyl propanoate, $CH_3CH_2CO_2C_6H_5$		211			i
Methyl benzoate, $C_6H_5CO_2CH_3$	-12.3	196	1.089	1.517	i
Ethyl benzoate, $C_6H_5CO_2CH_2CH_3$	-34.6	213	1.047	1.506	i
Propyl benzoate, $C_6H_5CO_2(CH_2)_2CH_3$	-51.6	231	1.023	1.500	i
1-Methylethyl benzoate, $C_6H_5CO_2CH(CH_3)_2$		218	1.011		i
Butyl benzoate, $C_6H_5CO_2(CH_2)_3CH_3$	-22.4	247	1.000	1.497	i
Phenyl benzoate, $C_6H_5CO_2C_6H_5$	71	314			i
Phenylmethyl benzoate, $C_6H_5CO_2CH_2C_6H_5$	19.4	327	1.112/ 25 °C	1.568	i
AMIDES					
Methanamide, $HCONH_2$	2.6	193 dec.	1.134	1.445	vs
Ethanamide, CH_3CONH_2	82.3	221			s
Propanamide, $CH_3CH_2CONH_2$	81.3	213			s
Butanamide, $CH_3(CH_2)_2CONH_2$	116	216			s
Benzamide, $C_6H_5CONH_2$	132	290			{ vs(hot) ss
Urea, $CO(NH_2)_2$	133	dec.	1.335	1.484	vs
Ethanediamide, $H_2NOCCONH_2$	419 dec.				i
NITRILES AND ISONITRILES					
Methanonitrile, HCN	-14	26			vs
Ethanonitrile, CH_3CN	-45.7	81.6	0.784	1.346	vs
Ethanoisonitrile, CH_3NC	-45	60	0.76		
Propanonitrile, CH_3CH_2CN	-91.9	97.2	0.772	1.363	vs
Propanoisonitrile, CH_3CH_2NC		79	0.76		
Butanonitrile, $CH_3(CH_2)_2CN$	-112	118	0.791	1.384	ss
2-Methylpropanonitrile, $(CH_3)_2CHCN$		108		1.372	
Benzonitrile, C_6H_5CN	-13	191	1.006	1.529	ss(hot)
Benzoisonitrile, C_6H_5NC		165–166 dec.	0.98		
AMINES					
Methylamine, CH_3NH_2	-92.5	-6.3			vs
Ethylamine, $CH_3CH_2NH_2$	-84	16.6			vs
Propylamine, $CH_3(CH_2)_2NH_2$	-83	48.6	0.719	1.388	vs
Butylamine, $CH_3(CH_2)_3NH_2$	-50.5	77.8	0.741	1.401	vs
Pentylamine, $CH_3(CH_2)_4NH_2$	-55	103	0.761	1.413	vs
Hexylamine, $CH_3(CH_2)_5NH_2$	-19	130	0.768	1.425	ss
Cyclohexylamine, $\underline{CH_2(CH_2)_4CHNH_2}$	-17.7	134	0.867	1.459	s
Phenylmethylamine, $C_6H_5CH_2NH_2$		184	0.981	1.540	vs
Phenylamine (aniline), $C_6H_5NH_2$	-6.2	184	1.022	1.586	ss
2-Methylphenylamine, $CH_3C_6H_4NH_2$	-16.2	200	0.999	1.569	ss
3-Methylphenylamine, $CH_3C_6H_4NH_2$	-43.6	203	0.992	1.569	ss
4-Methylphenylamine, $CH_3C_6H_4NH_2$	43.5	200			ss

Name and formula	$\theta_{C,m}/°C$	$\theta_{C,b}/°C$	$\rho/g\ cm^{-3}$	n_D	solu-bility
Dimethylamine, $(CH_3)_2NH$	−96	7·4			vs
Trimethylamine, $(CH_3)_3N$	−117	3·5			vs
Diethylamine, $(CH_3CH_2)_2NH$	−48	56·3	0·707	1·386	vs
Triethylamine, $(CH_3CH_2)_3N$	−115	89·4	0·728	1·400	⎰ss(hot) ⎱vs(< 18°C)
Diphenylamine, $(C_6H_5)_2NH$	52·8	302			i
Triphenylamine, $(C_6H_5)_3N$	126	365			i
N-Methylphenylamine, $C_6H_5NHCH_3$	−57	196	0·989	1·571	i
NN-Dimethylphenylamine, $C_6H_5N(CH_3)_2$	2·5	194	0·956	1·558	ss

NITRO-COMPOUNDS

Nitromethane, CH_3NO_2	−28·5	101	1·136	1·394	s
Nitroethane, $CH_3CH_2NO_2$	−90	115	1·052	1·392	s
1-Nitropropane, $CH_3(CH_2)_2NO_2$	−108	131	1·003	1·402	ss
1-Nitrobutane, $CH_3(CH_2)_3NO_2$		153		1·410	ss
Nitrobenzene, $C_6H_5NO_2$	5·7	211	1·204	1·556	i
Methyl-2-nitrobenzene, $CH_3C_6H_4NO_2$	−2·9	220	1·163	1·546	i
Methyl-3-nitrobenzene, $CH_3C_6H_4NO_2$	15	233	1·157	1·547	i
Methyl-4-nitrobenzene, $CH_3C_6H_4NO_2$	51·7	238			i
1,2-Dinitrobenzene, $C_6H_4(NO_2)_2$	118	319			ss
1,3-Dinitrobenzene, $C_6H_4(NO_2)_2$	90·0	291			i
1,4-Dinitrobenzene, $C_6H_4(NO_2)_2$	72	299			i
Methyl-2,4-dinitrobenzene, $CH_3C_6H_3(NO_2)_2$	70	300 dec.			i
Methyl-2,4,6-trinitrobenzene $CH_3C_6H_2(NO_2)_3$	82	240 expl.			i

57 Melting Temperatures of Organic Derivatives

$\theta_{C,m}$ denotes melting temperature and dec. denotes decomposes.

(a) Derivatives of Alcohols

	3,5-Dinitro-benzoate $\theta_{C,m}/°C$		3,5-Dinitro-benzoate $\theta_{C,m}/°C$
Methanol	109	2-Methylpropan-2-ol	142
Ethanol	94	Pentan-1-ol	46
Propan-1-ol	75	Hexan-1-ol	61
Propan-2-ol	122	Phenylmethanol	113
Butan-1-ol	64	Cyclohexanol	113
2-Methylpropan-1-ol	88	Ethane-1,2-diol (glycol)	169[a]
Butan-2-ol	76		

[a] Disubstituted derivative

(b) Derivatives of Phenols

	3,5-Dinitro-benzoate $\theta_{C,m}/°C$	4-Methyl-benzene-sulphonate $\theta_{C,m}/°C$		3,5-Dinitro-benzoate $\theta_{C,m}/°C$	4-Methyl-benzene-sulphonate $\theta_{C,m}/°C$
Phenol	146	96	Benzene-1,2-diol	152[a]	
2-Methylphenol	138	55	Benzene-1,3-diol	201[a]	81[a]
3-Methylphenol	165	51	Benzene-1,4-diol	317[a]	159[a]
4-Methylphenol	189	70	2-Nitrophenol	155	83
Naphthalen-1-ol	217	88	3-Nitrophenol	159	113
Naphthalen-2-ol	210	125	4-Nitrophenol	188	97

(c) Derivatives of Aldehydes and Ketones

	2,4-Dinitro-phenyl-hydrazone $\theta_{C,m}/°C$		2,4-Dinitro-phenyl-hydrazone $\theta_{C,m}/°C$
Methanal	166	Propanone	126
Ethanal	168	Butanone	116
Propanal	155	Pentan-3-one	156
Butanal	126	Pentan-2-one	144
Benzaldehyde	237	Heptan-4-one	75
2-Hydroxybenzaldehyde	252dec.	Phenylethanone	250
Ethanedial	327	Diphenylmethanone	239
Trichloroethanal	131	Cyclohexanone	162

(d) Derivatives of Amines

	Ethanoyl derivative $\theta_{C,m}/°C$	Benzoyl derivative $\theta_{C,m}/°C$	4-Methyl-benzene-sulphonyl derivative $\theta_{C,m}/°C$
Methylamine	28	80	75
Ethylamine	205[b]	69	62
Propylamine	47	85	52
Butylamine	229[b]	70	65
(Phenylmethyl)amine	60	105	116
Phenylamine	114	163	103
Cyclohexylamine	104	147	87
2-Methylphenylamine	112	143	110
3-Methylphenylamine	66	125	114
4-Methylphenylamine	152	158	118
Dimethylamine	116[b]	42	87
Diethylamine	186[b]	42	60
Diphenylamine	103	180	142

[a] Disubstituted derivative
[b] Boiling temperature

58 Strengths of Organic Acids

The acid strengths in the following tables are given in terms of pK_a values, where $pK_a = -\log_{10}(K_a/mol\,dm^{-3})$. The dissociation constants K_a are for aqueous solutions at 298 K.

(a) Simple Aliphatic and Aromatic Acids

Name	Formula	pK_a
Methanoic	HCO_2H	3·75
Ethanoic	CH_3CO_2H	4·76
Propanoic	$CH_3CH_2CO_2H$	4·87
Butanoic	$CH_3(CH_2)_2CO_2H$	4·82
2-Methylpropanoic	$(CH_3)_2CHCO_2H$	4·85
Pentanoic	$CH_3(CH_2)_3CO_2H$	4·86
Trimethylethanoic	$(CH_3)_3CCO_2H$	5·05
Benzoic	$C_6H_5CO_2H$	4·20
Phenylethanoic	$C_6H_5CH_2CO_2H$	4·31
Propenoic	$CH_2=CHCO_2H$	4·26

(b) Halogen-Substituted Aliphatic Acids

Name	Formula	pK_a
Ethanoic	CH_3CO_2H	4·76
Chloroethanoic	CH_2ClCO_2H	2·86
Dichloroethanoic	$CHCl_2CO_2H$	1·29
Trichloroethanoic	CCl_3CO_2H	0·65
Fluoroethanoic	CH_2FCO_2H	2·66
Bromoethanoic	CH_2BrCO_2H	2·90
Iodoethanoic	CH_2ICO_2H	3·17
Propanoic	$CH_3CH_2CO_2H$	4·87
2-Chloropropanoic	$CH_3CHClCO_2H$	2·83
3-Chloropropanoic	$CH_2ClCH_2CO_2H$	4·10
Butanoic	$CH_3CH_2CH_2CO_2H$	4·82
2-Chlorobutanoic	$CH_3CH_2CHClCO_2H$	2·84
3-Chlorobutanoic	$CH_3CHClCH_2CO_2H$	4·06
4-Chlorobutanoic	$CH_2ClCH_2CH_2CO_2H$	4·52

(c) Other Substituted Aliphatic Acids

Name	Formula	pK_a
Ethanoic	CH_3CO_2H	4·76
Hydroxyethanoic (glycollic)	CH_2OHCO_2H	3·83
Aminoethanoic (glycine)	$CH_2NH_2CO_2H$	9·87
Cyanoethanoic	CH_2CNCO_2H	2·47
Methoxyethanoic	$CH_3OCH_2CO_2H$	3·53
3-Oxobutanoic	$CH_3COCH_2CO_2H$	3·58

(d) Substituted Aromatic Acids

Name	Formula	pK_a
Benzoic	$C_6H_5CO_2H$	4·20
2-Chlorobenzoic	$ClC_6H_4CO_2H$	2·94
3-Chlorobenzoic	$ClC_6H_4CO_2H$	3·83
4-Chlorobenzoic	$ClC_6H_4CO_2H$	3·99
2-Hydroxybenzoic (salicylic)	$HOC_6H_4CO_2H$	2·99
3-Hydroxybenzoic	$HOC_6H_4CO_2H$	4·08
4-Hydroxybenzoic	$HOC_6H_4CO_2H$	4·58
2-Aminobenzoic	$H_2NC_6H_4CO_2H$	6·97
3-Aminobenzoic	$H_2NC_6H_4CO_2H$	4·78
4-Aminobenzoic	$H_2NC_6H_4CO_2H$	4·92
2-Methylbenzoic	$H_3CC_6H_4CO_2H$	3·91
3-Methylbenzoic	$H_3CC_6H_4CO_2H$	4·24
4-Methylbenzoic	$H_3CC_6H_4CO_2H$	4·34
2-Nitrobenzoic	$O_2NC_6H_4CO_2H$	2·17
3-Nitrobenzoic	$O_2NC_6H_4CO_2H$	3·45
4-Nitrobenzoic	$O_2NC_6H_4CO_2H$	3·43
cis-Phenylpropenoic	$C_6H_5CH=CHCO_2H$	3·89
trans-Phenylpropenoic	$C_6H_5CH=CHCO_2H$	4·44

(e) Dicarboxylic Acids

Name	Formula	$pK_{a,1}$	$pK_{a,2}$
Ethanedioic	HO_2CCO_2H	1·23	4·28
Propanedioic	$HO_2CCH_2CO_2H$	2·83	5·69
Butanedioic	$HO_2C(CH_2)_2CO_2H$	4·22	5·64
Pentanedioic	$HO_2C(CH_2)_3CO_2H$	4·34	5·41
Hexanedioic	$HO_2C(CH_2)_4CO_2H$	4·43	5·28
Benzene-1,2-dicarboxylic	$HO_2CC_6H_4CO_2H$	2·98	5·41
Benzene-1,3-dicarboxylic	$HO_2CC_6H_4CO_2H$	3·46	4·60
Benzene-1,4-dicarboxylic	$HO_2CC_6H_4CO_2H$	3·51	4·82
cis-Butenedioic	$HO_2CCH=CHCO_2H$	1·92	6·23
trans-Butenedioic	$HO_2CCH=CHCO_2H$	3·02	4·38

(f) Phenols

Name	Formula	pK_a
Phenol	C_6H_5OH	10·00
2-Chlorophenol	ClC_6H_4OH	8·48
3-Chlorophenol	ClC_6H_4OH	9·02
4-Chlorophenol	ClC_6H_4OH	9·38
2,4,6-Trichlorophenol	$Cl_3C_6H_2OH$	7·6
Benzene-1,2-diol (catechol)	HOC_6H_4OH	9·85
Benzene-1,3-diol (resorcinol)	HOC_6H_4OH	9·81
Benzene-1,4-diol (hydroquinone)	HOC_6H_4OH	10·35
2-Methylphenol	$H_3CC_6H_4OH$	10·28
3-Methylphenol	$H_3CC_6H_4OH$	10·08
4-Methylphenol	$H_3CC_6H_4OH$	10·26
2-Nitrophenol	$O_2NC_6H_4OH$	7·21
3-Nitrophenol	$O_2NC_6H_4OH$	8·35
4-Nitrophenol	$O_2NC_6H_4OH$	7·15
2,4-Dinitrophenol	$(O_2N)_2C_6H_3OH$	4·01
2,4,6-Trinitrophenol	$(O_2N)_3C_6H_2OH$	0·42

(g) Alcohols

Name	Formula	pK_a
Methanol	CH_3OH	15·5
Ethanol	C_2H_5OH	ca. 16

The data in the following tables refer to the conjugate acid BH^+ of the base B named in the first column:

$$\underset{base}{B} + H_3O^+ = \underset{\substack{conjugate \\ acid}}{BH^+} + H_2O$$

(h) Primary Amines

Name	Formula	pK_a
Ammonia	NH_3	9·25
Methylamine	CH_3NH_2	10·64
Ethylamine	$CH_3CH_2NH_2$	10·73
Propylamine	$CH_3(CH_2)_2NH_2$	10·84
Butylamine	$CH_3(CH_2)_3NH_2$	10·61
Phenylamine	$C_6H_5NH_2$	4·62
Naphthalen-1-amine	$C_{10}H_7NH_2$	3·92
Naphthalen-2-amine	$C_{10}H_7NH_2$	4·11
Phenylmethylamine	$C_6H_5CH_2NH_2$	9·37
Cyclohexylamine	$C_6H_{11}NH_2$	10·64

(i) Secondary and Tertiary Amines

Name	Formula	pK_a
Dimethylamine	$(CH_3)_2NH$	10·72
Trimethylamine	$(CH_3)_3N$	9·80
Diethylamine	$(C_2H_5)_2NH$	10·93
Triethylamine	$(C_2H_5)_3N$	10·64
N-Methylphenylamine	$C_6H_5NHCH_3$	4·40
NN-Dimethylphenylamine	$C_6H_5N(CH_3)_2$	4·38
Diphenylamine	$(C_6H_5)_2NH$	0·8
Triphenylamine	$(C_6H_5)_3N$	very strong

(j) Substituted Aromatic Amines

Name	Formula	pK_a
Phenylamine	$C_6H_5NH_2$	4·62
2-Chlorophenylamine	$ClC_6H_4NH_2$	2·56
3-Chlorophenylamine	$ClC_6H_4NH_2$	3·46
4-Chlorophenylamine	$ClC_6H_4NH_2$	3·93
2-Hydroxyphenylamine	$HOC_6H_4NH_2$	4·72
3-Hydroxyphenylamine	$HOC_6H_4NH_2$	4·17
4-Hydroxyphenylamine	$HOC_6H_4NH_2$	5·47
2-Methylphenylamine	$H_3CC_6H_4NH_2$	4·38
3-Methylphenylamine	$H_3CC_6H_4NH_2$	4·67
4-Methylphenylamine	$H_3CC_6H_4NH_2$	5·00
2-Nitrophenylamine	$O_2NC_6H_4NH_2$	−0·28
3-Nitrophenylamine	$O_2NC_6H_4NH_2$	2·45
4-Nitrophenylamine	$O_2NC_6H_4NH_2$	0·98
N-Phenylethanamide	$C_6H_5NHCOCH_3$	0·61

(k) Amides

Name	Formula	pK_a
Ethanamide	CH_3CONH_2	−1·1
Urea	$CO(NH_2)_2$	0·18

59 Dipole Moments of Organic Compounds

Two equal and opposite electric charges of magnitude Q separated by a distance r produce a dipole moment μ defined by $\mu = Q \times r$. Many molecules have a permanent dipole moment: the debye, symbol D, is used as a unit for this physical quantity ($1D \cong 3.335\,640 \times 10^{-30}$ C m).

(a) Hydrocarbons

Compound	μ/D	Compound	μ/D
Alkanes (unbranched)	0	Cyclopropane	0
Ethene	0	Benzene	0
Propene	0·35	Methylbenzene	0·36
But-1-ene	0·38	Ethylbenzene	0·59
Ethyne	0	1,2-Dimethylbenzene	0·62
Propyne	0·75	1,4-Dimethylbenzene	0

(b) Halogen Derivatives

Compound	μ/D	Compound	μ/D
Chloromethane	1·87	Chlorobenzene	1·70
Bromomethane	1·81	Bromobenzene	1·64
Iodomethane	1·62	Iodobenzene	1·42
Dichloromethane	1·54	1,2-Dichlorobenzene	2·52
Trichloromethane	1·02	1,3-Dichlorobenzene	1·72
Tetrachloromethane	0	1,4-Dichlorobenzene	0
Chloroethane	2·06	(Chloromethyl)benzene	1·85
Bromoethane	2·02	cis-1,2-Dichloroethene	1·90
Iodoethane	1·90	trans-1,2-Dichloroethene	0
1,1-Dichloroethane	2·06	Chloroethene	1·45
1,2-Dichloroethane	1·19	1-Chloroprop-2-ene	1·90

(c) Other Aliphatic Compounds

Compound	μ/D	Compound	μ/D
Methanol	1·71	Ethanoic anhydride	2·8
Ethanol	1·70	Ethanoyl chloride	2·72
Propan-1-ol	1·69	Methyl ethanoate	1·72
Propan-2-ol	1·60	Ethyl ethanoate	1·78
Ethane-1,2-diol	2·28	Ethanamide	3·6
Methoxymethane	1·30	Urea	4·56
Methoxyethane	1·23	Ethanonitrile	3·84
Ethoxyethane	1·15	Propanonitrile	4·00
Epoxyethane	1·90	Nitromethane	3·44
Methanal	2·27	Nitroethane	3·54
Ethanal	2·72	Ethyl nitrite	2·38
Propanal	2·72	Methylamine	1·26
Propanone	2·88	Ethylamine	1·22
Methanoic acid	1·52	Propylamine	1·17
Ethanoic acid	1·74	Dimethylamine	1·03
Propanoic acid	1·75	Trimethylamine	0·67

(d) Other Benzene Derivatives

Compound	μ/D	Compound	μ/D
Phenylmethanol	1·67	Phenylamine	1·53
Phenol	1·45	N-Methylphenylamine	1·68
Methoxybenzene	1·38	Nitrobenzene	4·27
Ethoxybenzene	1·45	1,2-Dinitrobenzene	6·00
Phenoxybenzene	1·23	1,3-Dinitrobenzene	3·89
Benzaldehyde	2·76	1,4-Dinitrobenzene	0
Phenylethanone	3·02	2-Nitrophenylamine	4·24
Diphenylmethanone	2·95	3-Nitrophenylamine	4·94
Benzenesulphonic acid	3·8	4-Nitrophenylamine	6·2
Ethyl benzoate	2·00	Benzene-1,2-diamine	1·53
Benzamide	3·6	Benzene-1,3-diamine	1·81
Benzonitrile	4·42	Benzene-1,4-diamine	1·53

60 Standard Solutions for Titrimetric Analysis

Substance	Relative molecular mass
Acid-base titrations	
Benzoic acid, $C_6H_5CO_2H$	122·12
Hydrochloric acid, HCl	36·46
Aminosulphonic acid, NH_2SO_3H	97·09
Disodium tetraborate (borax), $Na_2B_4O_7.10H_2O$	381·37
Sodium carbonate, Na_2CO_3	105·99
Redox titrations	
Ammonium iron(II) sulphate, $(NH_4)_2Fe(SO_4)_2.6H_2O$	392·14
Arsenic(III) oxide, As_2O_3	197·84
Sodium ethanedioate, $Na_2C_2O_4$	134·00
Iodine, I_2	253·81
Potassium dichromate(VI), $K_2Cr_2O_7$	294·20
Potassium iodate(V), KIO_3	214·01
Precipitation titrations	
Silver, Ag	107·87
Silver nitrate, $AgNO_3$	169·87
Sodium chloride, NaCl	58·44
Complexometric titrations	
edta (disodium salt, dihydrate), $Na_2H_2C_{10}H_{12}O_8N_2.2H_2O$	372·24

61 Acid-base Indicators

The pK_{in} value of an indicator is defined by $pK_{in} = -\log_{10}(K_{in}/\text{mol dm}^{-3})$ where K_{in} denotes the indicator constant.

Indicator	pK_{in}	pH *range*	Colour change Acid	Alkali
Thymol blue	1·7	1·2–2·8	Red	Yellow
Methyl orange	3·7	3·1–4·4	Red	Yellow
Bromophenol blue	4·0	3·0–4·6	Yellow	Blue
Bromocresol green	4·7	3·8–5·4	Yellow	Blue
Methyl red	5·1	4·2–6·3	Red	Yellow
Bromothymol blue	7·0	6·0–7·6	Yellow	Blue
Phenol red	7·9	6·8–8·4	Yellow	Red
Thymol blue (2nd range)	8·9	8·0–9·6	Yellow	Blue
Phenolphthalein	9·3	8·3–10·0	Colourless	Red
Thymolphthalein	9·7	9·3–10·5	Colourless	Blue

62 Standard Buffer Solutions

The buffer solutions in the following table are suitable for purposes for which an accurate knowledge of pH is required, e.g. in the calibration of pH meters.

m denotes molality.

Buffer solution	pH *at* 298 K
Potassium hydrogen 2,3-dihydroxybutanedioate (tartrate), saturated at 298 K	3·557
Potassium hydrogen benzene-1,2-dicarboxylate (phthalate), $m = 0·05$ mol kg^{-1}	4·008
KH_2PO_4, $m = 0·025$ mol kg^{-1} Na_2HPO_4, $m = 0·025$ mol kg^{-1}	6·865
$Na_2B_4O_7$, $m = 0·01$ mol kg^{-1}	9·180

63 Laboratory Reagents

c denotes concentration.

Substance	Density/ gm cm^{-3}	Approximate per cent by mass	c mol dm^{-3}
Ammonia solution, concentrated	0·88	28	16
Ethanoic acid, glacial	1·05	99·5	17
Hydrobromic acid, concentrated	1·49	48	9
Hydrochloric acid, concentrated	1·18	36	12
Hydriodic acid, concentrated	1·70	55	7
Nitric acid, concentrated	1·42	70	16
Nitric acid, fuming	1·50	95	24
Phosphoric(V) acid, concentrated	1·75	85	15
Sulphuric acid, concentrated	1·84	98	18
Sulphuric acid, fuming	1·92	(20% SO$_3$)	

64 Infrared Data

The following table gives values for characteristic infrared absorptions due to stretching in organic molecules. Wavenumber $= \dfrac{1}{\text{wavelength}}$.

Bond		Wavelength/μm	Wavenumber/cm^{-1}
C−H	alkanes	3·38–3·51	2850–2960
	alkenes	3·23–3·32	3010–3095
	alkynes	3·03–3·08	3250–3300
	arenes	3·25–3·30	3030–3080
C−C		9·09–13·33	750–1100
C=C		5·95–6·17	1620–1680
C≡C		4·44–4·76	2100–2250
C⋯C	arenes	6·25–6·67	1500–1600
C−O	alcohols, ethers, carboxylic acids, esters	7·69–10·00	1000–1300
C=O	aldehydes, ketones, carboxylic acids, esters	5·71–5·95	1680–1750
C−N	amines	7·35–8·47	1180–1360
C≡N	nitriles	4·42–4·52	2210–2260
O−H	non-hydrogen-bonded (alcohols, phenols)	2·72–2·79	3580–3670
	hydrogen-bonded (alcohols, phenols)	2·82–3·10	3230–3550
	hydrogen-bonded (carboxylic acids)	3·33–4·00	2500–3000
N−H	non-hydrogen-bonded (amines, amides)	2·81–3·01	3320–3560
	hydrogen-bonded (primary and secondary amines)	2·94–3·23	3100–3400
C−Cl		12·5–16·7	600–800
C−Br		16·7–20·0	500–600
C−I		20·0	500

65 Composition of the Atmosphere

Gas	Volume per cent in dry air	Gas	Volume per cent in dry air
N_2	78·09	He	0·00052
O_2	20·95	Kr	0·00011
Ar	0·93	H_2	0·00005
CO_2	0·03	Xe	0·000009
Ne	0·0018	Rn	6×10^{-18}

66 Composition of the Earth's Crust

Note. The earth's crust includes the terrestrial waters and the atmosphere.

Element	Mass per cent	Element	Mass per cent
Oxygen	49·5	Chlorine	0·19
Silicon	25·7	Phosphorus	0·12
Aluminium	7·5	Carbon	0·09
Iron	4·7	Manganese	0·08
Calcium	3·4	Barium	0·05
Sodium	2·6	Sulphur	0·05
Potassium	2·4	Chromium	0·03
Magnesium	1·9	Fluorine	0·03
Hydrogen	0·88	Nitrogen	0·03
Titanium	0·58	All others	0·15

LOGARITHMS OF NUMBERS

	0	1	2	3	4	5	6	7	8	9	Differences								
											1	2	3	4	5	6	7	8	9
10	0000	0043	0086	0128	0170	0212					4	8	13	17	21	25	29	34	38
						0212	0253	0294	0334	0374	4	8	12	16	20	24	28	32	36
11	0414	0453	0492	0531	0569	0607					4	8	12	16	19	23	27	31	35
						0607	0645	0682	0719	0755	4	7	11	15	18	22	26	30	33
12	0792	0828	0864	0899	0934	0969					4	7	11	14	18	21	25	28	32
						0969	1004	1038	1072	1106	3	7	10	14	17	20	24	27	31
13	1139	1173	1206	1239	1271	1303					3	7	10	13	16	20	23	26	30
						1303	1335	1367	1399	1430	3	6	10	13	16	19	22	26	29
14	1461	1492	1523	1553	1584	1614	1644	1673	1703	1732	3	6	9	12	15	18	21	24	27
15	1761	1790	1818	1847	1875	1903	1931	1959	1987	2014	3	6	8	11	14	17	20	22	25
16	2041	2068	2095	2122	2148	2175	2201	2227	2253	2279	3	5	8	10	13	16	18	21	23
17	2304	2330	2355	2380	2405	2430	2455	2480	2504	2529	2	5	7	10	12	15	17	20	22
18	2553	2577	2601	2625	2648	2672	2695	2718	2742	2765	2	5	7	9	12	14	16	19	21
19	2788	2810	2833	2856	2878	2900	2923	2945	2967	2989	2	4	7	9	11	13	15	18	20
20	3010	3032	3054	3075	3096	3118	3139	3160	3181	3201	2	4	6	8	11	13	15	17	19
21	3222	3243	3263	3284	3304	3324	3345	3365	3385	3404	2	4	6	8	10	12	14	16	18
22	3424	3444	3464	3483	3502	3522	3541	3560	3579	3598	2	4	6	8	10	11	13	15	17
23	3617	3636	3655	3674	3692	3711	3729	3747	3766	3784	2	4	6	7	9	11	13	15	17
24	3802	3820	3838	3856	3874	3892	3909	3927	3945	3962	2	4	5	7	9	11	13	14	16
25	3979	3997	4014	4031	4048	4065	4082	4099	4116	4133	2	3	5	7	9	10	12	14	15
26	4150	4166	4183	4200	4216	4232	4249	4265	4281	4298	2	3	5	6	8	10	11	13	14
27	4314	4330	4346	4362	4378	4393	4409	4425	4440	4456	2	3	5	6	8	9	11	13	14
28	4472	4487	4502	4518	4533	4548	4564	4579	4594	4609	2	3	5	6	8	9	11	12	14
29	4624	4639	4654	4669	4683	4698	4713	4728	4742	4757	1	3	4	6	7	9	10	12	13
30	4771	4786	4800	4814	4829	4843	4857	4871	4886	4900	1	3	4	6	7	8	10	11	13
31	4914	4928	4942	4955	4969	4983	4997	5011	5024	5038	1	3	4	5	7	8	10	11	12
32	5051	5065	5079	5092	5105	5119	5132	5145	5159	5172	1	3	4	5	7	8	9	11	12
33	5185	5198	5211	5224	5237	5250	5263	5276	5289	5302	1	3	4	5	7	8	9	10	12
34	5315	5328	5340	5353	5366	5378	5391	5403	5416	5428	1	3	4	5	6	8	9	10	11
35	5441	5453	5465	5478	5490	5502	5514	5527	5539	5551	1	2	4	5	6	7	8	10	11
36	5563	5575	5587	5599	5611	5623	5635	5647	5658	5670	1	2	4	5	6	7	8	10	11
37	5682	5694	5705	5717	5729	5740	5752	5763	5775	5786	1	2	3	5	6	7	8	9	10
38	5798	5809	5821	5832	5843	5855	5866	5877	5888	5899	1	2	3	5	6	7	8	9	10
39	5911	5922	5933	5944	5955	5966	5977	5988	5999	6010	1	2	3	4	6	7	8	9	10
40	6021	6031	6042	6053	6064	6075	6085	6096	6107	6117	1	2	3	4	5	6	7	9	10
41	6128	6138	6149	6160	6170	6180	6191	6201	6212	6222	1	2	3	4	5	6	7	8	9
42	6232	6243	6253	6263	6274	6284	6294	6304	6314	6325	1	2	3	4	5	6	7	8	9
43	6335	6345	6355	6365	6375	6385	6395	6405	6415	6425	1	2	3	4	5	6	7	8	9
44	6435	6444	6454	6464	6474	6484	6493	6503	6513	6522	1	2	3	4	5	6	7	8	9
45	6532	6542	6551	6561	6571	6580	6590	6599	6609	6618	1	2	3	4	5	6	7	8	9
46	6628	6637	6646	6656	6665	6675	6684	6693	6702	6712	1	2	3	4	5	6	7	7	8
47	6721	6730	6739	6749	6758	6767	6776	6785	6794	6803	1	2	3	4	5	5	6	7	8
48	6812	6821	6830	6839	6848	6857	6866	6875	6884	6893	1	2	3	4	5	5	6	7	8
49	6902	6911	6920	6928	6937	6946	6955	6964	6972	6981	1	2	3	4	4	5	6	7	8

The tables of Logarithms and Anti-logarithms are reproduced, with slight amendments, from Knott's *Four Figure Mathematical Tables*, by permission of the publishers, Messrs. W. and R. Chambers Ltd.

LOGARITHMS OF NUMBERS

	0	1	2	3	4	5	6	7	8	9	1 2 3	4 5 6	7 8 9
												Differences	
50	6990	6998 7007 7016			7024 7033 7042			7050 7059 7067			1 2 3	3 4 5	6 7 8
51	7076	7084 7093 7101			7110 7118 7126			7135 7143 7152			1 2 3	3 4 5	6 7 8
52	7160	7168 7177 7185			7193 7202 7210			7218 7226 7235			1 2 2	3 4 5	6 7 7
53	7243	7251 7259 7267			7275 7284 7292			7300 7308 7316			1 2 2	3 4 5	6 6 7
54	7324	7332 7340 7348			7356 7364 7372			7380 7388 7396			1 2 2	3 4 5	6 6 7
55	7404	7412 7419 7427			7435 7443 7451			7459 7466 7474			1 2 2	3 4 5	5 6 7
56	7482	7490 7497 7505			7513 7520 7528			7536 7543 7551			1 2 2	3 4 5	5 6 7
57	7559	7566 7574 7582			7589 7597 7604			7612 7619 7627			1 2 2	3 4 5	5 6 7
58	7634	7642 7649 7657			7664 7672 7679			7686 7694 7701			1 2 2	3 4 4	5 6 7
59	7709	7716 7723 7731			7738 7745 7752			7760 7767 7774			1 1 2	3 4 4	5 6 7
60	7782	7789 7796 7803			7810 7818 7825			7832 7839 7846			1 1 2	3 4 4	5 6 6
61	7853	7860 7868 7875			7882 7889 7896			7903 7910 7917			1 1 2	3 4 4	5 6 6
62	7924	7931 7938 7945			7952 7959 7966			7973 7980 7987			1 1 2	3 3 4	5 6 6
63	7993	8000 8007 8014			8021 8028 8035			8041 8048 8055			1 1 2	3 3 4	5 6 6
64	8062	8069 8075 8082			8089 8096 8102			8109 8116 8122			1 1 2	3 3 4	5 5 6
65	8129	8136 8142 8149			8156 8162 8169			8176 8182 8189			1 1 2	3 3 4	5 5 6
66	8195	8202 8209 8215			8222 8228 8235			8241 8248 8254			1 1 2	3 3 4	5 5 6
67	8261	8267 8274 8280			8287 8293 8299			8306 8312 8319			1 1 2	3 3 4	4 5 6
68	8325	8331 8338 8344			8351 8357 8363			8370 8376 8382			1 1 2	3 3 4	4 5 6
69	8388	8395 8401 8407			8414 8420 8426			8432 8439 8445			1 1 2	3 3 4	4 5 6
70	8451	8457 8463 8470			8476 8482 8488			8494 8500 8506			1 1 2	2 3 4	4 5 6
71	8513	8519 8525 8531			8537 8543 8549			8555 8561 8567			1 1 2	2 3 4	4 5 5
72	8573	8579 8585 8591			8597 8603 8609			8615 8621 8627			1 1 2	2 3 4	4 5 5
73	8633	8639 8645 8651			8657 8663 8669			8675 8681 8686			1 1 2	2 3 4	4 5 5
74	8692	8698 8704 8710			8716 8722 8727			8733 8739 8745			1 1 2	2 3 4	4 5 5
75	8751	8756 8762 8768			8774 8779 8785			8791 8797 8802			1 1 2	2 3 3	4 5 5
76	8808	8814 8820 8825			8831 8837 8842			8848 8854 8859			1 1 2	2 3 3	4 5 5
77	8865	8871 8876 8882			8887 8893 8899			8904 8910 8915			1 1 2	2 3 3	4 4 5
78	8921	8927 8932 8938			8943 8949 8954			8960 8965 8971			1 1 2	2 3 0	4 4 5
79	8976	8982 8987 8993			8998 0004 9009			9015 9020 9025			1 1 2	2 3 3	4 4 5
80	9031	9036 9042 9047			9053 9058 9063			9069 9074 9079			1 1 2	2 3 3	4 4 5
81	9085	9090 9096 9101			9106 9112 9117			9122 9128 9133			1 1 2	2 3 3	4 4 5
82	9138	9143 9149 9154			9159 9165 9170			9175 9180 9186			1 1 2	2 3 3	4 4 5
83	9191	9196 9201 9206			9212 9217 9222			9227 9232 9238			1 1 2	2 3 3	4 4 5
84	9243	9248 9253 9258			9263 9269 9274			9279 9284 9289			1 1 2	2 3 3	4 4 5
85	9294	9299 9304 9309			9315 9320 9325			9330 9335 9340			1 1 2	2 3 3	4 4 5
86	9345	9350 9355 9360			9365 9370 9375			9380 9385 9390			1 1 2	2 3 3	4 4 5
87	9395	9400 9405 9410			9415 9420 9425			9430 9435 9440			0 1 1	2 2 3	3 4 4
88	9445	9450 9455 9460			9465 9469 9474			9479 9484 9489			0 1 1	2 2 3	3 4 4
89	9494	9499 9504 9509			9513 9518 9523			9528 9533 9538			0 1 1	2 2 3	3 4 4
90	9542	9547 9552 9557			9562 9566 9571			9576 9581 9586			0 1 1	2 2 3	3 4 4
91	9590	9595 9600 9605			9609 9614 9619			9624 9628 9633			0 1 1	2 2 3	3 4 4
92	9638	9643 9647 9652			9657 9661 9666			9671 9675 9680			0 1 1	2 2 3	3 4 4
93	9685	9689 9694 9699			9703 9708 9713			9717 9722 9727			0 1 1	2 2 3	3 4 4
94	9731	9736 9741 9745			9750 9754 9759			9763 9768 9773			0 1 1	2 2 3	3 4 4
95	9777	9782 9786 9791			9795 9800 9805			9809 9814 9818			0 1 1	2 2 3	3 4 4
96	9823	9827 9832 9836			9841 9845 9850			9854 9859 9863			0 1 1	2 2 3	3 4 4
97	9868	9872 9877 9881			9886 9890 9894			9899 9903 9908			0 1 1	2 2 3	3 4 4
98	9912	9917 9921 9926			9930 9934 9939			9943 9948 9952			0 1 1	2 2 3	3 4 4
99	9956	9961 9965 9969			9974 9978 9983			9987 9991 9996			0 1 1	2 2 2	3 3 4

ANTILOGARITHMS

	0	1	2	3	4	5	6	7	8	9	Differences								
											1	2	3	4	5	6	7	8	9
·00	1000	1002	1005	1007	1009	1012	1014	1016	1019	1021	0	0	1	1	1	1	1	2	2
·01	1023	1026	1028	1030	1033	1035	1038	1040	1042	1045	0	0	1	1	1	1	2	2	2
·02	1047	1050	1052	1054	1057	1059	1062	1064	1067	1069	0	0	1	1	1	1	2	2	2
·03	1072	1074	1076	1079	1081	1084	1086	1089	1091	1094	0	0	1	1	1	1	2	2	2
·04	1096	1099	1102	1104	1107	1109	1112	1114	1117	1119	0	1	1	1	1	2	2	2	2
·05	1122	1125	1127	1130	1132	1135	1138	1140	1143	1146	0	1	1	1	1	2	2	2	2
·06	1148	1151	1153	1156	1159	1161	1164	1167	1169	1172	0	1	1	1	1	2	2	2	2
·07	1175	1178	1180	1183	1186	1189	1191	1194	1197	1199	0	1	1	1	1	2	2	2	2
·08	1202	1205	1208	1211	1213	1216	1219	1222	1225	1227	0	1	1	1	1	2	2	2	3
·09	1230	1233	1236	1239	1242	1245	1247	1250	1253	1256	0	1	1	1	1	2	2	2	3
·10	1259	1262	1265	1268	1271	1274	1276	1279	1282	1285	0	1	1	1	1	2	2	2	3
·11	1288	1291	1294	1297	1300	1303	1306	1309	1312	1315	0	1	1	1	2	2	2	2	3
·12	1318	1321	1324	1327	1330	1334	1337	1340	1343	1346	0	1	1	1	2	2	2	2	3
·13	1349	1352	1355	1358	1361	1365	1368	1371	1374	1377	0	1	1	1	2	2	2	3	3
·14	1380	1384	1387	1390	1393	1396	1400	1403	1406	1409	0	1	1	1	2	2	2	3	3
·15	1413	1416	1419	1422	1426	1429	1432	1435	1439	1442	0	1	1	1	2	2	2	3	3
·16	1445	1449	1452	1455	1459	1462	1466	1469	1472	1476	0	1	1	1	2	2	2	3	3
·17	1479	1483	1486	1489	1493	1496	1500	1503	1507	1510	0	1	1	1	2	2	2	3	3
·18	1514	1517	1521	1524	1528	1531	1535	1538	1542	1545	0	1	1	1	2	2	2	3	3
·19	1549	1552	1556	1560	1563	1567	1570	1574	1578	1581	0	1	1	1	2	2	3	3	3
·20	1585	1589	1592	1596	1600	1603	1607	1611	1614	1618	0	1	1	1	2	2	3	3	3
·21	1622	1626	1629	1633	1637	1641	1644	1648	1652	1656	0	1	1	2	2	2	3	3	3
·22	1660	1663	1667	1671	1675	1679	1683	1687	1690	1694	0	1	1	2	2	2	3	3	3
·23	1698	1702	1706	1710	1714	1718	1722	1726	1730	1734	0	1	1	2	2	2	3	3	4
·24	1738	1742	1746	1750	1754	1758	1762	1766	1770	1774	0	1	1	2	2	2	3	3	4
·25	1778	1782	1786	1791	1795	1799	1803	1807	1811	1816	0	1	1	2	2	2	3	3	4
·26	1820	1824	1828	1832	1837	1841	1845	1849	1854	1858	0	1	1	2	2	3	3	3	4
·27	1862	1866	1871	1875	1879	1884	1888	1892	1897	1901	0	1	1	2	2	3	3	3	4
·28	1905	1910	1914	1919	1923	1928	1932	1936	1941	1945	0	1	1	2	2	3	3	4	4
·29	1950	1954	1959	1963	1968	1972	1977	1982	1986	1991	0	1	1	2	2	3	3	4	4
·30	1995	2000	2004	2009	2014	2018	2023	2028	2032	2037	0	1	1	2	2	3	3	4	4
·31	2042	2046	2051	2056	2061	2065	2070	2075	2080	2084	0	1	1	2	2	3	3	4	4
·32	2089	2094	2099	2104	2109	2113	2118	2123	2128	2133	0	1	1	2	2	3	3	4	4
·33	2138	2143	2148	2153	2158	2163	2168	2173	2178	2183	1	1	2	2	3	3	4	4	5
·34	2188	2193	2198	2203	2208	2213	2218	2223	2228	2234	1	1	2	2	3	3	4	4	5
·35	2239	2244	2249	2254	2259	2265	2270	2275	2280	2286	1	1	2	2	3	3	4	4	5
·36	2291	2296	2301	2307	2312	2317	2323	2328	2333	2339	1	1	2	2	3	3	4	4	5
·37	2344	2350	2355	2360	2366	2371	2377	2382	2388	2393	1	1	2	2	3	3	4	4	5
·38	2399	2404	2410	2415	2421	2427	2432	2438	2443	2449	1	1	2	2	3	3	4	4	5
·39	2455	2460	2466	2472	2477	2483	2489	2495	2500	2506	1	1	2	2	3	3	4	5	5
·40	2512	2518	2523	2529	2535	2541	2547	2553	2559	2564	1	1	2	2	3	4	4	5	5
·41	2570	2576	2582	2588	2594	2600	2606	2612	2618	2624	1	1	2	2	3	4	4	5	5
·42	2630	2636	2642	2649	2655	2661	2667	2673	2679	2685	1	1	2	2	3	4	4	5	6
·43	2692	2698	2704	2710	2716	2723	2729	2735	2742	2748	1	1	2	3	3	4	4	5	6
·44	2754	2761	2767	2773	2780	2786	2793	2799	2805	2812	1	1	2	3	3	4	4	5	6
·45	2818	2825	2831	2838	2844	2851	2858	2864	2871	2877	1	1	2	3	3	4	5	5	6
·46	2884	2891	2897	2904	2911	2917	2924	2931	2938	2944	1	1	2	3	3	4	5	5	6
·47	2951	2958	2965	2972	2979	2985	2992	2999	3006	3013	1	1	2	3	3	4	5	5	6
·48	3020	3027	3034	3041	3048	3055	3062	3069	3076	3083	1	1	2	3	4	4	5	6	6
·49	3090	3097	3105	3112	3119	3126	3133	3141	3148	3155	1	1	2	3	4	4	5	6	6

ANTILOGARITHMS

	0	1	2	3	4	5	6	7	8	9	Differences 1 2 3	4 5 6	7 8 9
·50	3162	3170	3177	3184	3192	3199	3206	3214	3221	3228	1 1 2	3 4 4	5 6 7
·51	3236	3243	3251	3258	3266	3273	3281	3289	3296	3304	1 2 2	3 4 5	5 6 7
·52	3311	3319	3327	3334	3342	3350	3357	3365	3373	3381	1 2 2	3 4 5	5 6 7
·53	3388	3396	3404	3412	3420	3428	3436	3443	3451	3459	1 2 2	3 4 5	6 6 7
·54	3467	3475	3483	3491	3499	3508	3516	3524	3532	3540	1 2 2	3 4 5	6 6 7
·55	3548	3556	3565	3573	3581	3589	3597	3606	3614	3622	1 2 2	3 4 5	6 7 7
·56	3631	3639	3648	3656	3664	3673	3681	3690	3698	3707	1 2 3	3 4 5	6 7 8
·57	3715	3724	3733	3741	3750	3758	3767	3776	3784	3793	1 2 3	3 4 5	6 7 8
·58	3802	3811	3819	3828	3837	3846	3855	3864	3873	3882	1 2 3	4 4 5	6 7 8
·59	3890	3899	3908	3917	3926	3936	3945	3954	3963	3972	1 2 3	4 5 5	6 7 8
·60	3981	3990	3999	4009	4018	4027	4036	4046	4055	4064	1 2 3	4 5 6	6 7 8
·61	4074	4083	4093	4102	4111	4121	4130	4140	4150	4159	1 2 3	4 5 6	7 8 9
·62	4169	4178	4188	4198	4207	4217	4227	4236	4246	4256	1 2 3	4 5 6	7 8 9
·63	4266	4276	4285	4295	4305	4315	4325	4335	4345	4355	1 2 3	4 5 6	7 8 9
·64	4365	4375	4385	4395	4406	4416	4426	4436	4446	4457	1 2 3	4 5 6	7 8 9
·65	4467	4477	4487	4498	4508	4519	4529	4539	4550	4560	1 2 3	4 5 6	7 8 9
·66	4571	4581	4592	4603	4613	4624	4634	4645	4656	4667	1 2 3	4 5 6	7 9 10
·67	4677	4688	4699	4710	4721	4732	4742	4753	4764	4775	1 2 3	4 6 7	8 9 10
·68	4786	4797	4808	4819	4831	4842	4853	4864	4875	4887	1 2 3	4 6 7	8 9 10
·69	4898	4909	4920	4932	4943	4955	4966	4977	4989	5000	1 2 3	5 6 7	8 9 10
·70	5012	5023	5035	5047	5058	5070	5082	5093	5105	5117	1 2 4	5 6 7	8 9 11
·71	5129	5140	5152	5164	5176	5188	5200	5212	5224	5236	1 2 4	5 6 7	8 10 11
·72	5248	5260	5272	5284	5297	5309	5321	5333	5346	5358	1 2 4	5 6 7	9 10 11
·73	5370	5383	5395	5408	5420	5433	5445	5458	5470	5483	1 3 4	5 6 8	9 10 11
·74	5495	5508	5521	5534	5546	5559	5572	5585	5598	5610	1 3 4	5 6 8	9 10 12
·75	5623	5636	5649	5662	5675	5689	5702	5715	5728	5741	1 3 4	5 7 8	9 10 12
·76	5754	5768	5781	5794	5808	5821	5834	5848	5861	5875	1 3 4	5 7 8	9 11 12
·77	5888	5902	5916	5929	5943	5957	5970	5984	5998	6012	1 3 4	5 7 8	10 11 12
·78	6026	6039	6053	6067	6081	6095	6109	6124	6138	6152	1 3 4	6 7 8	10 11 13
·79	6166	6180	6194	6209	6223	6237	6252	6266	6281	6295	1 3 4	6 7 9	10 11 13
·80	6310	6324	6339	6353	6368	6383	6397	6412	6427	6442	1 3 4	6 7 9	10 12 13
·81	6457	6471	6486	6501	6516	6531	6546	6561	6577	6592	2 3 5	6 8 9	11 12 14
·82	6607	6622	6637	6653	6668	6683	6699	6714	6730	6745	2 3 5	6 8 9	11 12 14
·83	6761	6776	6792	6808	6823	6839	6855	6871	6887	6902	2 3 5	6 8 9	11 13 14
·84	6918	6934	6950	6966	6982	6998	7015	7031	7047	7063	2 3 5	6 8 10	11 13 14
·85	7079	7096	7112	7129	7145	7161	7178	7194	7211	7228	2 3 5	7 8 10	12 13 15
·86	7244	7261	7278	7295	7311	7328	7345	7362	7379	7396	2 3 5	7 9 10	12 14 15
·87	7413	7430	7447	7464	7482	7499	7516	7534	7551	7568	2 3 5	7 9 10	12 14 16
·88	7586	7603	7621	7638	7656	7674	7691	7709	7727	7745	2 4 5	7 9 11	12 14 16
·89	7762	7780	7798	7816	7834	7852	7870	7889	7907	7925	2 4 5	7 9 11	13 14 16
·90	7943	7962	7980	7998	8017	8035	8054	8072	8091	8110	2 4 6	7 9 11	13 15 17
·91	8128	8147	8166	8185	8204	8222	8241	8260	8279	8299	2 4 6	8 10 11	13 15 17
·92	8318	8337	8356	8375	8395	8414	8433	8453	8472	8492	2 4 6	8 10 12	14 15 17
·93	8511	8531	8551	8570	8590	8610	8630	8650	8670	8690	2 4 6	8 10 12	14 16 18
·94	8710	8730	8750	8770	8790	8810	8831	8851	8872	8892	2 4 6	8 10 12	14 16 18
·95	8913	8933	8954	8974	8995	9016	9036	9057	9078	9099	2 4 6	8 10 12	15 17 19
·96	9120	9141	9162	9183	9204	9226	9247	9268	9290	9311	2 4 6	8 11 13	15 17 19
·97	9333	9354	9376	9397	9419	9441	9462	9484	9506	9528	2 4 7	9 11 13	15 17 20
·98	9550	9572	9594	9616	9638	9661	9683	9705	9727	9750	2 4 7	9 11 13	15 18 20
·99	9772	9795	9817	9840	9863	9886	9908	9931	9954	9977	2 5 7	9 11 14	16 18 21

Index

Acid-base indicators, 103
Activation energy, 86
Antilogarithms, 108
Atmosphere, composition, 105
Atomic mass, relative, 8
Atomic number, 3, 4, 8
Atomic spectrum of hydrogen, 23
Azeotropes, 64

Boiling temperature, elements, 8, 50
    organic compounds, 87
Bond angle, 30
Bond enthalpy, average, 32
Bond length, covalent, 31
Buffer solutions, 103

Complex ions, stability constant, 77
Conductivity, electric, elements, 18
    electrolytic, potassium chloride, 79
    electrolytic, water, 81
    molar, aqueous solutions, 80
    ions, 80
Constants, physical, 1
Conversion factors, 1
Covalent radius, 27
Crystal structure of elements, 12

Density, elements, 8
    organic compounds, 87
Dipole moment, inorganic compounds, 26
    organic compounds, 100
Dissociation constant, inorganic acids, 74
    organic acids, 96

Earth's crust, composition, 105
Electric conductivity of elements, 18
Electrode, standard reference, 73
Electrode potential standard, 70
Electromagnetic spectrum, 2
Electron affinity, 22
Electronegativity, 24
Electronic configuration of elements, 4
Ellingham diagrams, 53
Energy levels of many-electron atom, 7

Enthalpy, atomization, 50
    bond, average, 32
    combustion, 54
    formation, aqueous ions, 57
    elements, 50
    inorganic compounds, 33
    organic compounds, 48
    hydration, 58
    hydrogenation, 55
    lattice, 59
    melting, compounds, 51
    neutralization, 56
    solution, 56
    vaporization, compounds, 51
    elements, 50
Entropy, elements, 33
    inorganic compounds, 33
    organic compounds, 48
Equilibrium constant, gaseous reactions, 68
Equilibrium data, 67
Eutectic mixtures, 66

Fundamental particles, 1

Gibbs free energy, formation, aqueous ions, 57
    inorganic compounds, 33
    organic compounds, 48
    metal extraction, 53

Half-life, radioactive, 16, 17
Heat capacity, molar, elements, 33
    inorganic compounds, 33
    organic compounds, 48
    specific, elements, 8
Hybrid orbitals, 31
Hydration enthalpy, 58

Indicators, acid-base, 103
Infrared data, 104
Ionic character, percentage, 25
Ionic conductivity, 80
Ionic product of water, 75
Ionic radius, 28
Ionization energy, 19

Isotopes, half-life, 16, 17
   stable, abundance, 13

Kinetic data, 82

Laboratory reagents, 104
Lattice enthalpy, 59
Logarithms, 106

Melting temperature, elements, 8, 50
   organic compounds, 87
   organic derivatives, 94
Metallic radius, 27
Molar conductivity, aqueous solutions, 80
   ions, 80

Oxidation state, 8

Periodic classification, 3
Physical constants, 1

Radioactive series, 17
Radius, covalent, 27
   ionic, 28
   metallic, 27
   van der Waals, 27

Rate laws, 85
Refractive index, organic compounds, 87
Relative atomic mass, 8

Shapes of molecules and ions, 30
SI units, xiii
Solubility, gases, 60
   inorganic compounds, 62
   organic compounds, 87
Solubility product, 76
Specific heat capacity of elements, 8
Spectrum, atomic, hydrogen, 23
   electromagnetic, 2
Stability constant of complex ions, 77
Standard solutions, 102

Thermodynamic data, change of state, 50
   elements, 33
   inorganic compounds, 33
   metal extraction, 53
   organic compounds, 48
Transition temperature, 66

van der Waals radius, 27
Vapour pressure of water, 65